The Spy Archive

Hidden Lives, Secret Missions, and the History of Espionage

Dexter Ingram

Table of Contents

To my parents — thank you for opening the world to me, nurturing my love of travel, and instilling the value of service to our nation's security.

To my wife, Karima — your strength carried me through the year I spent in Afghanistan and every challenge before and after. I am forever grateful for your patience, courage, and love.

To our sons, Karim and Amir — thank you for embracing the adventures of our family's journey, for meeting new places and languages with curiosity and resilience.

This book is, in many ways, ours.

Introduction

What if I told you history's biggest turning points weren't decided by the generals whose statues fill our parks, but by people whose names we'll never know? There's a hidden world where empires crumble through secret exchanges, covert meetings, and silent plots—not through open warfare. This is the dark theater of espionage. Forget knights on horseback or stiff diplomats with quill pens; picture instead the nervous messenger slipping past guards with a tiny note that could topple a kingdom.

Most history lovers know the glossy version—the grand battles, the peace treaties signed with pizzazz. But behind these stories is intelligence, deception, and covert manipulation. A spider's web spun by faceless agents whose secrets touch everything. They've tilted the scales at critical moments, but their influence has vanished into historical footnotes.

Digging into these dark corners changes everything. Once you see how spies have pulled strings behind the scenes, you'll realize you can't really understand our world without understanding them. History's greatest chess players weren't sitting on thrones. They were hiding in plain sight, moving pieces no one else could see.

Espionage has evolved over the years. Think about it. The ancient Persians and Romans had to build elaborate networks of people just to gather intel. Now, our agencies use satellites and hack computers instead. But the heart of it hasn't changed a bit. We're still just power-hungry humans desperate for information. The covert operations of our past changed maps and brought down governments. When you trace the evolution from hidden messages in ancient scrolls to modern keystroke monitoring, you can't help but see how yesterday's spy techniques connect to the threats we're facing tomorrow.

Hollywood paints spies as sexy, martini-sipping adventurers living on adrenaline and seduction. The truth is far less glamorous. Real spying means gut-wrenching moral choices and crushing personal sacrifices—elements rarely captured in two-hour thrillers. Actual intelligence work demands patience, constant paranoia, and blind faith in causes most could never understand.

This gap between fiction and reality matters because it colors how we view national security. Movies burn lasting myths into our brains, casting spies as either flawless heroes or mustache-twirling villains. This distorted lens affects serious discussions about surveillance, privacy, and security, influencing decisions that impact millions. Cutting through these Hollywood myths allows for smarter, more nuanced conversations about intelligence work.

Beyond the movie fantasy, these invisible hands have flipped the outcomes of wars, rigged elections, and sabotaged peace talks. They've redrawn national boundaries and overthrown governments. They've steered global affairs in ways traditional history books often overlook. Looking at these hidden interventions reveals how thoroughly espionage has carved the world we take for granted.

Look around at today's world—messy, confusing, full of conflicts that seem to come from nowhere. Want to know what's really going on? Dig into the historical impact of espionage. This stuff isn't just for history nerds or thriller fans—it's the decoder for today's headlines. Many of those global hotspots burning on your news feed started with some spy's report filed decades ago. Once you understand these old games, today's baffling politics suddenly make a weird kind of sense. You become a sharper citizen, able to see past the smoke and mirrors of international affairs.

Truth is, espionage has always been history's invisible puppet master, yanking strings behind the scenes while we watch the public performance. Its dark fingerprints are everywhere, if you know how to spot them. This book pulls you beyond those sanitized history textbooks to uncover the buried stories and the real reasons nations went to war, made peace, or betrayed allies. Only when you acknowledge these shadows can you grasp the forces that built our world—and still pull the strings today.

Chapter 1:

Early Intelligence Networks

The messenger dropped to his knees, sweat dripping from his brow after racing across the desert. Thutmose III, Egypt's warrior pharaoh, leaned forward on his golden throne. The court fell silent. What came next wasn't just news—it was priceless. Secret information about enemy positions was gathered by what historians now recognize as the world's first documented spy network.

Three thousand years before CIA analysts existed, pharaohs were getting intelligence briefings. And they weren't alone. From the Nile to the Tigris, ancient rulers had already figured out the game-changing value of knowing their enemies' secrets.

The First Spymasters and Mesopotamian Protocol

Thutmose III completely changed how ancient Egypt handled espionage. People call him the "Egyptian Napoleon" because, like the famous French general, he knew winning battles started with good information. What made him special? He turned spy work into a system that gave him an edge over enemies who never saw him coming.

Egypt's Eyes and Ears

The oldest documented spy mission is found in an inscription from ancient Egypt, dating back around 1500 BCE. It describes how Pharaoh Thutmose III gathered intelligence before the Battle of Megiddo. Facing a coalition of Canaanite rulers, Thutmose needed to know which mountain passes weren't heavily defended. The intelligence worked—the pharaoh chose an unexpected route and achieved total surprise, crushing his opponents (Mark, 2017).

This victory wasn't a one-off success. Egyptian rulers had already established a system for collecting intelligence. The "House of Life" served as both a religious temple and an early intelligence headquarters where scribes recorded information about neighboring lands, resources, and potential threats. Pharaohs dispatched trusted representatives to foreign courts as early diplomats. Their missions included what we would now call diplomatic reporting and spying.

The Egyptians understood that knowledge was power long before Francis Bacon coined the phrase. Their military campaigns were backed by research. Their diplomatic missions doubled as intelligence operations. The foundation for spying was already in place.

But Egyptian spies brought more than just information back to their pharaohs. Often, they delivered death.

Masters of Silent Elimination

Egyptian spy-assassins were the original two-for-one special. Why just learn about your enemy when you could remove them entirely?

The tomb paintings don't show this darker side. You won't find hieroglyphs boasting about assassination programs. But archaeological evidence tells a different story.

These ancient spy-assassins had a particular affinity for snake venom— a weapon that left minimal evidence and mimicked natural illness. Egypt's snake handlers were respected members of society who performed at religious ceremonies. Behind their public role, some played a more covert role.

Everyone knew Egypt had snake handlers for religious rituals. Nobody suspected these same experts were teaching spies how to milk cobras and vipers for targeted killings.

From Egypt to Assyria

While Egyptian pharaohs sent spies scrambling across foreign lands, the Assyrians were doing something even more revolutionary: creating the ancient world's version of a classified database.

Deep in the royal palaces of Nineveh, archaeologists found thousands of clay tablets that read exactly like modern intelligence reports. The Assyrians wrote down everything: where enemy troops moved, what rival kings planned, and even the personality weaknesses of foreign leaders they could exploit later.

For example, a psychological profile of an Elamite prince, dated about 700 B.C.E., was found. The Assyrians noticed he was quick to anger, fond of wine, and estranged from his brother—all weaknesses their agents could use to their advantage. It wasn't just random gossip. This was actionable intel.

This systematic approach paid off big time for the Assyrian kings, ruling a huge empire with threats bubbling up everywhere. Their armies could strike fast because they weren't starting from zero when trouble flared. They'd pull out the relevant clay tablets, spot patterns and weaknesses, then hit their enemies where it hurt most.

One tablet from King Esarhaddon's reign shows just how sharp they were. When a revolt threatened in the western provinces, his spies provided names of rebel leaders, details about their resources, and notes about which local officials might secretly support them. The tablet even suggested that rebel leaders might be bribed into betraying their comrades (University of Toronto, 2010).

Switch that clay tablet for a digital file, and any modern intelligence officer would recognize exactly what they were looking at.

The Spy Game Hasn't Changed Much

What's crazy is that ancient spy operations relied 100% on human sources—what the CIA now calls HUMINT. No satellites, no wiretaps, no social media monitoring, yet they built networks that kept empires standing for centuries.

These ancient spies worked deep undercover as merchants or travelers. They bribed officials for secrets, found sources in enemy courts, and devised clever ways to smuggle information home safely. Egyptian spies hid messages in walking sticks and false-bottomed jars—tricks not so different from hollow coins and modified everyday items Cold War spies would use thousands of years later.

Different tools, same tricks. Hide who you really are. Get close to people with secrets. Don't get caught sending messages home. Every spy in history has played by these rules, whether they served pharaohs or presidents.

Biblical and Greek Espionage Operations

The Bible describes what might be history's most famous spy mission. When Moses sent twelve men to check out Canaan, he wasn't just curious—he was running classic battlefield surveillance with a specific mission.

"Go up this way into the South," Moses ordered. He continued:

> See what the land is like: whether the people who dwell in it are strong or weak, few or many; whether the land they dwell in is good or bad; whether the cities they inhabit are like camps or strongholds; whether the land is rich or poor; and whether there are forests there or not. (New King James Bible, 1982, Numbers 13:17–20).

Moses even told them to bring back samples of local produce, which they did, returning with a massive cluster of grapes so big it took two men to carry it on a pole.

The mission ended with a split verdict. Ten spies yelled, "We are not able to go up against the people, for they are stronger than we." Only two—Joshua and Caleb—argued, "We can take them" (New King James Bible, 1982, Numbers 13:31–33).

The Israelites went with the majority report. They paid for it with forty years of wandering in the desert until a new generation decided to listen to the minority opinion.

Let's be honest, if your intelligence operation leads to a 40-year delay in achieving your goal, something went terribly wrong.

The Greek Twist

While biblical accounts show early spy missions, the Greeks turned espionage into an official government function, especially during the heated rivalry between Athens and Sparta.

The Spartans established what they called the Crypteia, which was their secret service. At first, they used these guerrilla soldiers to terrorize and control their helot slave population. Young Spartan men would hide during the day, then pop out at night to spy on helots and eliminate anyone who looked suspicious (Ross, 2012).

But during the Peloponnesian War, the Spartans realized these skills worked well against Athens, too. Soon, they were sending these trained agents to gather information about the Athenians' defenses and political decisions.

In his seminal work, *History of the Peloponnesian War*, Thucydides gives numerous accounts of espionage operations that influenced the conflict's outcome. He documents how, in 425 B.C.E., Spartan intelligence gathering allowed them to ambush Athenian forces at Pylos after discovering their planned naval movements. Thucydides also explains how both sides used proxenos—citizens who served foreign interests in their home city—as valuable intelligence assets who could move freely between territories under diplomatic protection.

Athens fought back with the first counterintelligence operations in history. Their port officials carefully checked foreigners entering the city, watching for potential spies. They created laws making it illegal to share military plans or shipbuilding details. Athenian agents became good at intercepting Spartan messages and catching enemy spies in the act.

Thucydides discusses a Spartan diplomatic meeting at Athens that was meant to build anti-Athenian alliances. This meeting refers to Spartan negotiations with Persian satraps in 412–411 B.C.E., which Thucydides details in Book VIII. Athenian intelligence networks learned about these negotiations, allowing Athens to counter Sparta's diplomatic efforts by sending their own envoys to Persia. Detecting the plot early, Athens temporarily disrupted Sparta's plans to secure Persian funding.

Thucydides also describes how intelligence failures could be catastrophic, as with the Athenian expedition to Sicily in 415 B.C.E. Poor intelligence about Sicilian defenses and political alliances contributed to Athens' disastrous defeat. Thucydides saw this as an important aspect of statecraft—the need for reliable information before committing to major military ventures.

The Athenians realized something monumental. Intelligence isn't just learning your enemy's secrets—it's stopping them from learning yours.

The Stick That Changed History

During this ancient spy-versus-spy contest, the Spartans developed a cryptographic tool so effective that its basic concept would still be taught in military academies 2,500 years later: the "scytale."

A Spartan commander would take a staff (wooden rod), wrap a strip of parchment or leather around it spirally, then write his message across the wrapped material. When unwrapped, the letters appeared jumbled and meaningless.

It was the simplicity that made it so great. Only someone with a rod of identical diameter could rewrap the strip and read the message correctly. When wrapped around the matching rod, the jumbled letters would suddenly align into readable text.

This was history's first transposition cipher. Instead of substituting different symbols or characters for the original text—as the later Caesar cipher would do—the scytale kept the original letters but scrambled their positions.

The military advantage was significant. Spartan officers could communicate securely over great distances without worrying that enemies would read their plans.

What often gets overlooked is how naturally the scytale fit into Spartan society. The wooden rod was deeply practical for a warrior society and could be disguised as a simple walking staff. Additionally, any Spartan commander could use it with minimal training as the cipher required no memorization of complicated codes.

Roman influence

If the Greeks made espionage official, the Romans made it professional. Their most famous spy service, the Frumentarii, started with the most boring job imaginable and evolved into something far more interesting.

The Frumentarii began as grain collectors. They were basically supply officers who ensured armies got fed. But think about what that job involved: traveling to different places, talking to locals, and keeping detailed records. Perfect spy skills.

Over time, these ordinary-looking supply officers evolved into the emperor's personal intelligence service. Working from their headquarters at the Castra Peregrina in Rome, they ran operations throughout the empire. Their original grain-collecting duties were the perfect cover for gathering information about foreign threats and internal enemies.

Emperor Hadrian used them brilliantly. Hadrian knew the private lives of his friends so well that they couldn't ask him for anything he didn't already know about. The emperor would casually drop details about his associates' personal affairs into conversation, letting them know his spies missed nothing.

This wasn't just security. It was power politics. Hadrian used intelligence as leverage. When people know you have their secrets, they tend to behave themselves (Internet History Sourcebooks Project, 2025).

But even Rome's system failed sometimes. The ambush of General Varus and his three legions in Germany's Teutoburg Forest happened because Roman intelligence missed warnings about Germanic tribal alliances. And despite numerous rumors circulating beforehand,

Rome's spies never uncovered the plot against Julius Caesar in time to save him from those fateful knife wounds.

These failures teach us that even the best systems have blind spots. And information that reaches decision-makers too late—or gets ignored—might as well never have been collected at all. The Romans learned this lesson the hard way, with Caesar bleeding out on the Senate floor.

The Caesar Cipher

But Julius Caesar left another, more positive legacy to the spy world: one of history's most famous encryption systems. The "Caesar Cipher" was a systematic way to protect sensitive military communications.

Caesar used what cryptographers call a "substitution cipher." He'd shift each letter in a message by a fixed number of positions in the alphabet. So, with a shift of three, "A" becomes "D," "B" becomes "E," and so on.

This meant Caesar could send written orders to his generals without worrying about enemies catching and reading them. A seized message would look like gibberish to anyone who didn't know the specific shift number.

The Frumentarii adopted and expanded Caesar's method. They used similar ciphers when reporting sensitive information about potential rebels, foreign agents, or even politically dangerous figures at Rome's court.

Caesar's simple cipher system eventually proved vulnerable to frequency analysis—a technique where codebreakers look at how often certain symbols appear, matching them to the known frequency of letters in a language. The letter "E," for instance, appears more frequently than "Z" in Latin and most European languages.

Roman cryptographers recognized this weakness. Later imperial ciphers used multiple shift values or substitution tables to secure communications. The Frumentarii became quite sophisticated at these methods, especially for messages about internal threats that could be read by high-level sympathizers.

The Caesar Cipher represents one of history's first systematic encryption methods. Though simple by modern standards, it formed the foundation for cryptographic developments that continued through medieval monasteries to the era of computer-based encryption.

The Byzantine Secret Weapon

When Rome collapsed, the eastern half—what we call the Byzantine Empire—kept Roman intelligence traditions alive while incorporating their own tricks. Surrounded by enemies and unable to match them in military strength, the Byzantines turned to superior intelligence as their survival strategy.

Constantinople became medieval spy central. The Byzantines mastered the art of the merchant-spy, gathering secrets while running legitimate trade throughout the Mediterranean. Their diplomats doubled as intelligence officers, accumulating information as carefully as they negotiated treaties.

The Byzantines took spying to new heights. They set up spy networks in rival courts, stole diplomatic messages, and devised clever strategies for analyzing what they learned. Their thousand-year survival after Rome fell wasn't luck—it was thanks largely to knowing their enemies' plans before those enemies did.

Emperor Justinian's takeover of Italy is a good example. Before sending in armies, he flooded the region with agents who mapped out Ostrogothic defenses. Then he identified potential local allies and stirred up divisions among his enemies. When his general Belisarius finally invaded, he knew exactly what he would face.

Byzantine spies actively controlled events through what we'd now call influence operations. They spread rumors, bribed officials, and manipulated perceptions. They understood that sometimes changing what an enemy believes can be more powerful than fighting them directly.

Reflections

From Thutmose to the Byzantine emperors, the story of ancient intelligence shows surprising continuity. Despite huge technological differences, the fundamental principles stayed consistent across thousands of years: humans provided the most reliable information, secure communication methods were a must, and intelligence that arrived too late or wasn't believed made no difference at all.

Medieval kings built on Byzantine methods. Renaissance city-states refined diplomatic intelligence. By Cardinal Richelieu's time in 17th-century France, spy services were starting to look recognizably modern.

Despite this modernization, ancient roots are still present in today's high-tech agencies. When CIA officers meet sources in safe houses, when NSA analysts verify the reliability of intercepts, or when briefers present findings to presidents, they're doing jobs that the first spies would recognize.

In other words, lessons learned by the first intelligence officers—pressed into clay, written on papyrus, whispered to kings—still inspire today's intelligence officers. Their principles have survived centuries of warfare, political upheaval, and technological revolution. The tools keep changing, but the game remains the same.

Chapter 2:

Medieval Intelligence Networks

Under a towering castle, a shadow moved in the moonlight. Not a knight. Not a nobleman. Just a man with dirty fingernails and unremarkable clothes carrying a kingdom's fate inside a loaf of bread. To the right people, that piece of paper, marked with cipher symbols, was worth more than gold. It was worth dying for.

Medieval castles weren't just fortresses of stone and mortar. They were the epicenters of hidden truths. While poets sang of brave knights and battles, power plays took place in silence, coded messages, and midnight meetings. Behind those imposing walls, spies gathered secrets that could overthrow monarchies or preserve dynasties.

The romantic vision of medieval warfare gets it all wrong. Sure, there were battles with knights and archers. But the wise kings won before anyone drew a sword. They won through intelligence.

Every part of a castle tells two stories.

Castle Intelligence

Medieval castles were the birthplaces of organized intelligence. Inside these stone walls, physical defenses teamed up with human information networks to create something revolutionary.

Think about a castle's design. Every feature had multiple uses. Those high windows were perfect for archers, but also ideal for watching who comes and goes in the village below. The multiple courtyards created security layers where visitors could be observed before reaching the inner sanctum.

Even castle acoustics played a role. Some were deliberately designed with hidden listening posts—tiny chambers or passages where servants could eavesdrop on important conversations.

In Nottingham Castle, they built a tunnel called Mortimer's Hole. This famous passage was used to capture the Earl of March, Roger Mortimer, in 1330. But before that, it served as a way for spies to enter and exit unseen, bypassing the main gates where everyone was watching (Secret Passage, 2025).

Castle-based systems laid the foundation for specialized intelligence. The Knights Templar didn't invent their spy networks out of nothing—they relied on generations of castle-based espionage.

The Medieval Intelligence Game

The feudal system wasn't only about lords providing knights for battle—it was medieval Europe's first structured intelligence network.

William the Conqueror's famous *Domesday Book* from 1086 seems like a dry tax record at first glance. Look closer. This massive survey documented exactly what resources existed in every corner of England: who owned what land, how many pigs each village kept, which areas could support cavalry, and so on. This book meant William knew England better than the English did (*Domesday Book*, n.d.).

The Church played an integral role in medieval intelligence games. Parish priests heard everyone's confessions—from peasants to nobles—while monks could read and write at a time when literacy was rare. They moved from court to court with "religious texts" that often concealed explosive state secrets, royal conspiracies, and kingdom-ending battle plans.

During Scotland's fight for independence, Edward I got his best intelligence on William Wallace from Dominican friars. A simple group of travelers plotted rebel attacks while pretending to spread the gospel. Wallace never suspected that these men of God were actually King Edward's eyes (William Wallace, 2025).

Medieval rulers were experts at "triangulating information," or comparing reports from multiple sources to separate fact from fiction. During the Hundred Years' War, English kings sent Welsh archers to spy on French positions. Those guys knew the terrain inside and out and could verify troop locations firsthand, providing intelligence that helped plan battles and saved countless lives (Evans, 2019).

Henry V employed a similar technique during the 1418–1419 siege of Rouen. He personally interrogated deserters from the city, but he wasn't gullible—he questioned different informants separately, checking their stories to catch lies. If a baker claimed the city had three months of grain but a soldier said they were down to their last week, someone was lying. Henry would keep questioning different sources until he found the truth, then adjust his siege tactics based on what he learned.

Modern intelligence agencies refer to this process as source validation. In medieval times, kings didn't have spy satellites or wiretaps, but they understood the cardinal rule: never trust a single source.

Outfoxing Enemies

Sometimes, the shrewdest move wasn't to fight at all. Duke William of Normandy (the same man who later conquered England) faced invaders in 1054 but avoided direct confrontation (William I "The Conqueror," 2018).

William's intelligence network tracked exactly where enemy forces were headed. Then he'd watch them, hiding in plain sight. Whenever they got close to a village, he'd evacuate it. When they tried to find food, they found empty storehouses. His spies watched their every move and reported back, letting William stay one step ahead.

This strategy drove the invaders crazy. They couldn't force William into battle where he might get crushed. His constant movement of food and valuables meant they couldn't sustain themselves by raiding. Eventually, the frustrated invaders left. William won without fighting a major battle.

This victory wasn't luck. It was calculated strategic deception based on superior intelligence. William knew where his enemies were, but they never knew where he was.

During the Wars of the Roses, Edward IV showed how intelligence could create military miracles. He could mobilize troops at shocking speed thanks to his communication network. When his rival Warwick thought Edward was confined in one part of England, Edward suddenly appeared with an army somewhere completely different (The Wars of the Roses, n.d.).

Edward's intelligence network provided him with real-time updates about enemy positions, safe routes, and where to find supporters.

While his enemies worked with outdated information, Edward had fresh intelligence arriving daily. This intensive network helped him move faster than anyone thought possible.

In one famous incident, Edward's enemies believed they had him trapped, but his men found an obscure river crossing. While Warwick's forces guarded the main routes, Edward slipped his entire army across this hidden route and appeared where nobody expected him.

Pure intelligence victory. The medieval rulers who lived longest weren't necessarily the ones with the biggest armies. They were the ones who invested in knowing things—where enemies were weak, which paths were open, which enemy lords might switch sides.

The Knights Templar Intelligence System

The Knights Templar brilliantly combined banking with espionage, creating one of history's first international intelligence operations. Their commanderies scattered across Europe and the Middle East were effective intelligence hubs disguised as religious houses.

The Templars' stroke of genius was their financial system. They created a medieval version of encrypted communications by embedding intelligence into banking transactions. A routine money transfer between Paris and Jerusalem could be linked to troop movements or political shifts.

The papal bull Omne Datum Optimum granted the Templars incredible autonomy, allowing them to operate independently of local rulers. They were able to gather and transmit intelligence outside of royal control (Omne Datum Optimum, n.d.).

Their houses became clearinghouses for information. Knowledge from distant lands was acquired, scrutinized, and transmitted. A Templar commander in England might receive intelligence from Cyprus faster than the king's own messengers could deliver it.

Roger de Flor's story in the Mediterranean during the 13th century shows the sometimes volatile nature of medieval intelligence. This former Templar admiral used information as skillfully as his sword, playing the dangerous game of Mediterranean politics.

De Flor's spies blended in with the port crowds as traders. By carefully placing agents across key harbors, he built a system that kept him ahead of enemies and allies alike. A ship bringing wool from Cyprus might carry news of Byzantine troop movements in its captain's memory.

But spy work is deadly. De Flor was assassinated at what he thought was a friendly dinner. His killers studied his methods well. They knew the best traps were hidden by ceremony and smiles.

The Ladies That Spied

Then there's Eleanor of Aquitaine—perhaps medieval Europe's greatest spy. While everyone watched men with swords, she built an army of women who could enter any castle's great hall without raising suspicion.

Eleanor's specialty was exploiting society's prejudices. Who would suspect elegant ladies-in-waiting of being master spies? Her agents moved through royal courts collecting secrets while seemingly just engaging in gossip.

Her methods were ingenious. A lady-in-waiting might drop a seemingly innocent love poem with battle plans. A traveling nun could carry messages sewn into the seams of her robe. A merchant's wife hosting dinner parties might extract valuable information from guests while appearing harmless.

Even when Eleanor was imprisoned by her husband, Henry II, for supporting their sons' rebellion, her network continued to operate.

Her system was brilliant. It didn't depend on her personal freedom. Her ladies kept their covers as gossips and companions while continuing to move information across Europe.

Medieval Cryptography

In a time when one wrong move could start a war or destroy a kingdom, keeping secrets was a matter of life and death. Medieval spies devised creative ways to exchange information securely.

Religious texts were perfect vehicles for hiding messages. A prayer book might have battle plans written in biblical references. For example, Psalm 23 might indicate coordinates for troop movements, or the seven seals in Revelation could point to meeting locations.

The genius was in its simplicity. Small adjustments to numbers or notes meant nothing to most readers but everything to those who knew the key. And who would suspect subversion in a bishop's prayer book?

Charlemagne's court in the 8th and 9th centuries was the intellectual center of medieval cryptography. It's where Alcuin of York transformed cryptography. Between Latin lessons and theological debates, he taught scribes to write codes that protected royal communications across the empire.

Medieval cryptography was born in those palace chambers. But monasteries soon began to develop code as well. Monks weren't copying religious texts—they were inventing ways to protect information.

These scholarly brothers brought intellectual firepower to cryptography, going beyond simple letter substitutions. They rigorously tested each method before entrusting it with valuable documents.

The Couriers Game

Getting secret messages to their destination needed exceptional couriers—perhaps the most dangerous job in medieval espionage.

They weren't just messengers. They were actors playing high-stakes roles where one mistake meant death.

Here's a typical scenario: a courier stops at a city gate. Her tattered coat hides royal secrets. She's been rehearsing this route for weeks. A note written on parchment and hidden inside a holster must reach its destination without being detected.

At sunset, she'll leave the bread near an oak tree marked with three small stones. By midnight, another agent will collect it, neither knowing the other's identity.

These dead drops turned ordinary places into secret mailboxes. A gap in a church wall, a loose brick on a bridge, or a specific grave in a country cemetery could become sites for secret exchanges.

The most talented couriers vanished into crowds. They played their parts perfectly—the wandering minstrel singing coded warnings, the peasant who practiced his limp for months, or the monk whose prayers hid battle plans.

Medieval spies based their trade on nerves and timing. Each successful delivery meant outwitting guards, avoiding bandits, and staying calm when questioned. It was these nameless agents who wrote the rules of espionage with their lives.

Reflections

Did you notice how the most dangerous people in a medieval court rarely needed steel at their side? Look at the ladies-in-waiting. They appear to be fixing each other's hair, sharing idle gossip. But in

Eleanor's court, they might run Europe's spy system. That "accidental" wine spill on the Spanish ambassador's sleeve? Perfect excuse to get close and overhear his private conversation.

Medieval spies turned everything into potential messages. A monk tending herb gardens might plant rows of rosemary in a way that communicated troop movements. A "simple" prayer book might have plans written between Psalms.

Even the castles themselves got in on it. Architects built secret listening channels into walls, allowing conversations in the great hall to be heard in hidden chambers elsewhere. Guards wasted time searching for hidden papers while real messages passed through dinner arrangements or across candles on a table.

They weaponized the mundane. These events may make you wonder about those stone-faced aristocrat portraits. I bet half of them were plotting something massive while the artist was at work.

Chapter 3:

Eastern Shadows—Espionage in

Ancient Asia

In the bustling markets of ancient Xi'an, whispers carried secrets as precious as silk and spices. Merchants traded not only goods but also news of distant lands, sharing insights that could tip the scales of power. Beneath the vibrant clamor of negotiation and barter, a silent war unfolded—one fought not with swords or arrows, but with cunning minds and covert operations. These were the invisible architects of history, masters of strategy who shaped the fate of empires without ever setting foot on the battlefield.

As travelers and traders moved along the Silk Road, they unwittingly wove a web of information that stretched across Asia. From the meticulous plans of Sun Tzu in China to the shadowy agents of feudal Japan, espionage was an art form, a crucial element in the grand chess game of dynasties and empires. In this chapter, we will journey through time to uncover the hidden networks and sophisticated strategies that underpinned significant historical shifts. By exploring the ingenious methods developed by Eastern civilizations, we'll gain insight into how intelligence outmaneuvered raw might, leaving an indelible mark on military and political history.

Sun Tzu's Spy Revolution

The spy who changed ancient combat wouldn't stand out in a crowd. He was a 50-year-old advisor to a Chinese war council who understood that battles could be won without drawing a sword. His weapon wasn't physical force but carefully gathered intelligence that could shatter an enemy's will before combat began.

This covert war was played by strange rules. Sun Tzu publicly praised honorable combat while his spy networks quietly dismantled enemy defenses from within. Genghis Khan openly condemned treachery while his spies infiltrated cities months before his horsemen showed up.

The boundaries between spying, espionage, and psychological warfare disappeared. The friendly shopkeeper who is calculating grain prices in your market? He was testing your strength, spreading rumors about Mongol invincibility, and figuring out which city officials he could bribe. Was he just a trader, or the first wave of an invasion force?

Here's what made ancient spycraft so dangerous: Your savviest intelligence agents could be turned against you. Sun Tzu's five-category classification system for spies accounted for this constant threat—the double agent serving as both the most valuable and most treacherous operative you could deploy.

When Sun Tzu wrote *The Art of War*, he revealed an intelligence framework far beyond what previous military thinkers had imagined. This system could decide conflicts before armies collided.

The Secret War Behind Ancient Battlefields

Imagine this scene: The merciless afternoon sun blazed overhead as the general shifted anxiously beside Sun Tzu.

In the distance, rising dust clouds signaled the advancing Chu forces— exactly where Sun Tzu had predicted.

"How did you know their battle plan?" the general whispered.

"We didn't just predict it," Sun Tzu replied with a slight smile. "We created it."

The enemy's advance was no coincidence. It stemmed from a calculated deception operation where false information was fed to enemy spies, luring Chu forces into a perfectly prepared ambush— without a single Wu soldier engaging in combat.

This fifth-century B.C.E. operation exemplified the refinement of intelligence warfare—a revolution in military thinking that transformed how ancient powers fought and conquered each other.

The shift from direct confrontation to Sun Tzu's intelligence-driven system marked the greatest change in war since bronze weapons replaced stone.

Let's look at how Sun Tzu's spy networks created new types of advantage—and how three key types of operatives changed warfare forever.

The Eyes and Ears

In 341 B.C.E., a traveling trader arrived in a village near Chu's border. What villagers didn't suspect was that this ordinary trader was secretly gathering intelligence for a planned Wu offensive. His findings stunned Wu commanders—precise details of garrison strength, supply routes, and a hidden path through a supposedly impassable forest.

The merchant represented Sun Tzu's first category: the local spy. While earlier spying focused on immediate battlefield conditions, Sun Tzu's local spies provided strategic insights into enemy territories, resources, and weaknesses.

After blending into border communities, these operatives studied terrain features, identified gaps in local defenses, and gauged public sentiment toward rulers. Their job wasn't just watching but actively manipulating their surroundings to prepare for future military actions.

What made local spies so effective was their authentic connection to their target areas. They spoke regional dialects, understood local customs, and had legitimate business or family ties that justified their presence. Their deep cover made them invisible.

For Sun Tzu, local spies served as both the advance wave of invasion and an early warning system for enemy movements. They could find alternate routes around heavily defended positions, report on defense tactics used by local forces, and turn seemingly impossible situations into strategic opportunities.

The impact could be devastating. A single skilled local spy could neutralize defensive advantages built over generations. Armies became useless when attackers knew about the unguarded mountain path, and ambush positions failed when enemy commanders received warnings of their locations.

Though rarely credited in historical accounts, local spies formed the backbone of successful campaigns throughout ancient China. They turned everyday civilian interactions into valuable intelligence assets, operating from market stalls, traveling caravans, and village gatherings.

It took extraordinary patience. Effective local spies often spent years creating their cover identities before providing truly valuable information. Sun Tzu's systematic use of local spies fundamentally changed how ancient powers understood and exploited enemy vulnerabilities.

The Hidden Threat

While local spies supplied geographical and demographic information, Sun Tzu recognized that truly decisive intelligence came from within an enemy's inner circle. His second category—internal spies—represented a higher-risk, higher-reward asset.

In 506 B.C.E., King Helü of Wu launched a campaign against the powerful Chu state. Despite Chu's superior numbers, Wu forces achieved a stunning victory at the Battle of Boju. What historical records often miss is the role of internal spies who had spent years infiltrating the Chu court. They provided critical information about command disputes and defensive gaps.

The idea was brilliantly simple. Instead of trying to overcome Chu's military advantages through force, Sun Tzu placed operatives among enemy advisors, servants, and officials—creating intelligence sources within the centers of power themselves.

The attackers gained access to sensitive planning discussions, battle preparations, and strategic deliberations. These internal spies revealed which commanders harbored resentment, which units suffered from low morale, and which defensive positions had been neglected.

The results were impressive. Through these carefully placed operatives, commanders acquired advanced knowledge of enemy movements, the exact timing of supply convoys, and details about command conflicts that could be exploited during battle. These spies often manipulated information flows, delaying certain warnings or giving credibility to false reports.

This wasn't random infiltration. This was a systematic compromise of enemy decision-making.

In 496 B.C.E., Wu forces targeted and eliminated prominent Chu commanders at the start of a major battle—a strike made possible only by internal spies who knew exactly where these officers would be positioned.

Internal spies show the evolution of intelligence gathering from general information to targeted surveillance. While standard spying might reveal an enemy's size and position, internal spies could provide actual battle plans and command intentions.

The economics of warfare had completely changed. By placing and maintaining a handful of well-positioned internal spies, commanders could neutralize the advantages of forces many times their size.

These operations also exposed a vulnerability in ancient warfare: the interconnectedness of command structures. Kingdoms that believed their security came from massive armies realized that they were only as secure as their most disgruntled official or ambitious military officer.

The Ultimate Sacrifice

For missions of extreme danger with minimal chance of survival, Sun Tzu deployed what he called "expendable spies"—operatives sent on one-way, high-priority missions.

Wei Li noticed something strange about the reported Chu troop movements he'd been sent to verify. Intelligence from other sources suggested a massive force congregating near the southern mountains, but local farmers mentioned nothing unusual. As an expendable spy, Wei Li's job was to go deep into enemy territory and confirm or disprove these reports, knowing he probably wouldn't return.

What Wei Li didn't realize was that his mission served multiple purposes—verifying intelligence but also testing enemy counterintelligence capabilities. If captured, his interrogation would reveal whether Chu forces knew about Wu's interest in the southern approach.

This part is the harshest of Sun Tzu's intelligence doctrine. Expendable spies were sacrificial operatives whose capture or death served strategic purposes beyond the immediate intelligence they gathered.

For the first time in warfare, intelligence planners designed operations expecting operative losses. How did Sun Tzu convince people to accept such dangerous assignments? Historical records suggest a mix of extreme rewards promised to families, coercion, and the opportunity for condemned criminals to redeem themselves.

The Perfect Intelligence System

From a commander's perspective, Sun Tzu's five-category classification system provided extraordinary advantages that previous military leaders could only imagine.

First, it provided complete coverage of the battlespace. Local spies supplied geographic and demographic intelligence, internal spies revealed enemy intentions and capabilities, double agents compromised enemy intelligence networks, expendable spies tested enemy defenses, and surviving spies (those who returned safely with critical information) brought back intelligence too sensitive to entrust to other channels.

Second, this system allowed intelligence verification through multiple sources. Information from one type of spy could be checked against another, creating a resilient intelligence system that is resistant to deception.

Third, the framework offered clear feedback loops. Intelligence failures could be analyzed based on which category of operations generated faulty information, allowing for systematic improvement rather than random adjustments.

Sun Tzu's intelligence doctrine allowed agents to hide in plain sight. The distinction between ordinary merchants, servants, or travelers and sophisticated intelligence agents became impossible for enemy security forces to detect.

The Chess Masters

No operative was more valuable—or dangerous—than the double agent. These chess masters of ancient espionage represented the pinnacle of counterintelligence operations.

Historical records reveal not just individual double agents but an entire ecosystem of counterintelligence activities. One particularly effective tactic involved tracking enemy spies, turning them, and using them to feed false information back to enemy commanders.

While internal spies provided valuable intelligence, double agents created strategic confusion and paralysis within enemy ranks. They delivered misinformation that seemed particularly credible because it came through an enemy's own trusted channel.

Chinese records from the Warring States period (c. 475–221 B.C.E.) document numerous battles where armies marched into carefully prepared ambushes because their own intelligence networks had been compromised by double agents. Military leaders defended these tactics as necessary for survival in an era of constant warfare. They credited double-agent operations with preserving smaller states that would otherwise have been overwhelmed by larger powers with superior military resources.

The double agent networks marked a turning point where intelligence and counterintelligence became inseparable. They showed that in ancient warfare, the limiting factor in military success was no longer numerical strength but information integrity.

These operations flipped the traditional military model. Rather than focusing exclusively on strengthening one's own forces, Sun Tzu emphasized compromising the enemy's decision-making process by manipulating their intelligence channels.

It's like poisoning the water supply rather than fighting each soldier individually.

Double agent operations triggered intense rivalry between states, strained diplomatic relations, and forced significant reforms to military command structures throughout ancient China.

Rulers became increasingly paranoid about the reliability of intelligence reports, sometimes executing messengers who brought in unwelcome news out of fear they might be enemy agents.

The Silent Warriors

The mist clings to the Kiso Valley as the merchant trudges along the mountain path. His weathered hands grip a walking staff, and his back bends under silk bolts destined for Kyoto's markets. Farmers nod respectfully as he passes—the same merchant they've seen for years.

Neither the farmers nor the checkpoint guards realize this "merchant" has never sold a single bolt of silk. He is Kato Danzo, one of the most accomplished *shinobi* of the 16th century. He moves undetected through enemy territory to gather intelligence that will decide the fate of provinces and lords.

This was the true art of the *shinobi*—not the black-clad assassins of Hollywood, but intelligence operatives who mastered the subtle skill of blending into plain sight while gathering information that changed history.

Birth of the Shadow Warriors

Spy networks in Japan didn't begin with mystical mountain clans but with practical necessity during Japan's Sengoku period (1467–1615), when the country fragmented into warring states, each controlled by ambitious daimyo (feudal lords).

Independence created the ideal environment for creating specialized intelligence services. Iga and Koga became Japan's spy academies, training operatives who sold their services to the highest bidders across Japan's fractured political landscape.

Their recruits weren't supernatural warriors but often came from lower-ranking samurai families or farmers with intimate knowledge of the mountains. What separated them from common soldiers was specialized training focused on information gathering rather than open combat.

The techniques they devised were revolutionary for medieval Japan but would be recognizable to any intelligence professional today: developing cover identities, setting up agent networks, creating private communication systems, and psychological manipulation of targets.

The Reality Behind the Myths

The black suit (*shinobi shōzoku*) so iconic in popular culture was actually one of their least-used disguises. Historical records show that *shinobi* more commonly posed as farmers, merchants, monks, performers, or beggars—roles that granted freedom of movement and access to information.

Their weapons weren't exotic ninja straight swords or massive shuriken throwing stars. Historical *shinobi* favored simple, concealable weapons that wouldn't reveal their true identities if they were found.

Even their famous acrobatic skills have been exaggerated. While physical training was important, *shinobi* weren't superhuman wall-climbers or water-walkers. They were highly trained observers who could map castle defenses, spot guard rotations, or notice weaknesses in enemy formations.

Intelligence Networks That Changed History

Shinobi were geniuses at setting up massive intelligence networks rather than relying on individual heroics. They built elaborate systems of informants, couriers, and safe houses across Japan that could rapidly collect and transmit information.

These networks proved devastating during Japan's constant wars. During the 1583 campaign against the Takeda clan, warlord Oda Nobunaga employed *shinobi* to map mountain passes, identify Takeda supply lines, and pinpoint weaknesses in castle defenses. This intelligence allowed him to launch strikes that ultimately destroyed the once-powerful Takeda.

Perhaps most impressive was the *shinobi* communication system. Historical records from the Edo period describe how Iga and Koga *shinobi* adopted a method called "*goshiki-zumi*" (five-colored ink) to conceal messages. They used special inks made from natural materials that only became visible under specific conditions, such as when heated or when certain chemicals were added. According to the military manual "Bansenshukai," compiled in 1676, *shinobi* also used pre-arranged signal fires and arranged everyday objects to communicate. While specific operations weren't well documented, it seems these methods allowed intelligence to move undetected, giving their employers huge strategic advantages.

Masters of Psychological Warfare

They were masters of what we now call "black propaganda." They would spread carefully worded rumors intended to sow distrust among commanders. Then they'd plant false information about imminent attacks, forcing enemies to exhaust themselves while preparing for non-existent threats.

The psychological dimension of *shinobi* work has been largely overlooked. Their greatest weapon wasn't the sword or poison but their power to manipulate perceptions. They understood that battles could be won or lost in the enemy's minds before any actual fighting began.

This psychological expertise extended to counterintelligence. *Shinobi* were frequently used to track enemy spies or test the loyalty of a Lord's followers through sneaky ruses and false-flag operations.

Technological Innovators

While their psychological and intelligence gathering skills were the backbone of *shinobi* operations, they also devised specialized equipment that gave them tactical advantages.

They were medieval Japan's Q Branch. They modified and invented tools specifically designed for covert operations.

Archaeological evidence and historical manuals describe innovative devices: lightweight folding boats for crossing moats, special climbing equipment for scaling castle walls, and early night vision techniques that used special eye drops to temporarily enhance one's ability to see in the dark.

They also pioneered specialized infiltration techniques. The *mizugumo* ("water spider") was a device for walking across the surfaces of moats—essentially, primitive snowshoes that distributed weight evenly. While not as magical as walking on water, *shinobi* were able to cross water barriers that would normally stop them.

Perhaps most interesting was their use of gunpowder for non-lethal purposes. While firearms revolutionized Japanese warfare, *shinobi* used gunpowder to create smoke bombs, flash-bang devices, and signal systems.

The Greatest Shinobi Operations

While many operations went undocumented (successful espionage often leaves little historical trace), several missions changed Japan's history.

In 1581, the daimyo Uesugi Kagekatsu sent *shinobi* operatives to infiltrate Oda Nobunaga's headquarters and gather intelligence on his battle plans. The information allowed Kagekatsu to predict and counter Nobunaga's strategies, saving his domain from defeat.

During the lead-up to the Battle of Sekigahara in 1600, which established the Tokugawa Shogunate that would rule Japan for over

250 years, *shinobi* networks supplied intelligence that helped Tokugawa Ieyasu identify which daimyo might betray him. During the battle, he used this intelligence to counter potential defections.

Perhaps most impressive was the infiltration of Osaka Castle during the 1615 siege that marked the final consolidation of Tokugawa power. *Shinobi* agents posing as construction workers located defensive positions, tracked weapon caches, and even sabotaged portions of the castle's defenses before the final assault.

While Japan's *shinobi* mastered espionage within their island nation, across the continent, another power had been perfecting intelligence on an imperial scale.

Genghis Khan's Spy Empire

Picture this scene: the Mongol spy crouched low in the tall grass, watching the Chinese fortress in the distance. For three weeks, he lived within bowshot of the walls, counting guard rotations, watching supply deliveries, and keeping track of which gates were poorly managed. Tonight, he would slip away to rendezvous with the arrow riders waiting thirty miles back, delivering information that was priceless to the Great Khan's generals.

Five hundred miles away, a silk merchant haggled in a market square in Samarkand. Between negotiations over fabric prices, he casually extracted details about city defenses, political tensions between local leaders, and water supply locations. No one questioned his interest—traders needed such information for their own security. None suspected he answered to Mongol paymasters who had bankrolled his entire caravan for his intelligence.

While their arrows darkened the sky, it was the Mongols' mastery of intelligence that truly conquered the world.

Capitalism Meets Espionage

The silk merchant straightened his cap as he neared Bukhara's city gates. His caravan had traveled the Silk Road for months, and his goods—colorful fabrics, spices, and luxury items—had gotten him entry to cities across Central Asia. The guards recognized him, greeted him warmly, and asked what news he brought from the East.

As he talked about carefully curated stories, he was already mentally cataloging the changes since his last visit: new defensive works on the eastern wall; fewer guards than before; and tension between the city governor and local military commander, evident from the cool reception when they met. All of these details would make their way back to Mongol intelligence handlers.

This commercial espionage operated on multiple levels. At the most basic level, ordinary merchants were debriefed when they returned to Mongol territories, and their commercial findings were repurposed for military planning. More elaborate operations involved merchants directly hired as agents and given specific intelligence requirements before they left.

The most valuable were the "dark" operatives—merchants who spent years setting up legitimate businesses in target cities, gaining credibility and access that could be activated when Mongol interest turned toward their region.

The Arrow Riders

While merchants provided human intelligence, the legendary "arrow riders" formed what might be called the world's first professional intelligence agency. Forget everything you think you know about Mongol warriors. These weren't just soldiers checking what was over the next hill. Arrow riders had specialized training in monitoring, memory techniques, and environmental analysis. They could estimate enemy numbers by looking at horse droppings, assess defense weaknesses at a glance, and memorize landscapes with extraordinary accuracy.

Their name came from their authorization tokens—arrow-shaped badges that granted them absolute priority on the Yam, the Mongol relay system. When an arrow rider arrived at a station, fresh horses would be prepped immediately, allowing non-stop movement of intelligence.

An arrow rider could cover 200 miles in a day by switching horses at stations spaced 20 miles apart. Information moved across the Mongol Empire faster than anywhere else on Earth at that time. Intelligence that would take weeks to reach European kings reached Mongol commanders in days.

Speed gave them an edge. During the 1241 invasion of Hungary, arrow riders marched months ahead of the main Mongol forces. They created detailed maps and assessed Hungarian defensive capabilities. By the time armies clashed at the Battle of Mohi, Mongol commanders had detailed intelligence about their opponents, their defenses, and the battlefield itself.

Arrow riders worked in specialized teams, each focusing on different facets of intelligence gathering. Some concentrated on military information: troop strengths, weapon types, or defense details. Others assessed economic factors: harvests, storage facilities, or livestock numbers. Then you had those who focused on political intelligence: finding tensions between leaders, potential collaborators, or resistance elements. The Hungarians never stood a chance.

Eastern vs. Western Traditions

Eastern philosophical thinking saw properly applied deception as a smart and virtuous tactic, and that idea shaped how their intelligence systems evolved. Sun Tzu's famous line, "all warfare is based on deception" (Tzu, n.d.), wasn't controversial—it reflected a cultural comfort with indirection that ran through Eastern warfare tactics.

Look at traditional Chinese thinking. The ideal victory wasn't beating your opponent in open battle. It was manipulating them into defeating themselves.

Direct confrontation? That was considered crude and wasteful. If you were really skilled as a general or spy, you won by making the enemy misread the situation until they made fatal mistakes.

This philosophical foundation created intelligence systems where deception wasn't just tolerated but preferred. Chinese imperial intelligence operations emphasized long-term penetration, patience, and psychological manipulation. Japanese ninja traditions similarly valued indirect methods, with agents spending years creating covers before acting.

Western traditions went in a completely different direction. European military philosophy, heavily influenced by Greek and Roman traditions, put direct confrontation on a pedestal as the honorable way to win.

Western military culture historically had a love affair with the frontal assault. Picture knights charging across open fields, or later, those infantry squares advancing in formation. Intelligence existed, sure, but it played second fiddle to battlefield heroics. The cultural ideal was facing your enemy directly—not working in the background.

This preference for direct confrontation meant early Western intelligence often lacked the sophistication of its Eastern counterparts. While Chinese emperors built elaborate spy networks that were integrated into their governance, European kings typically had no systematic approach, cobbling together arrangements with individual agents.

There was always this tension in Western intelligence. Espionage was necessary, but it was also dirty. The spy might serve the king, but he wasn't celebrated like the knight. This cultural discomfort with deception held back Western intelligence development for centuries.

Bureaucracy vs. Improvisation

The Chinese imperial system utilized sophisticated bureaucratic structures to gather intelligence centuries before anything similar appeared in the West. Eastern ideologies treated intelligence as fundamental to statecraft, creating permanent organizations with specialized functions.

The Han Dynasty founded the Yushi Tai—the Censorate—nearly 2,000 years ago. More than a spy agency, it gathered intelligence, monitored corruption, assessed public opinion, and reported directly to the emperor. The West had nothing even close until the 1800s.

By building strong bureaucracies, Eastern intelligence services were able to keep records and pass down skills over generations. The Chinese *Feng zheng* (secret police) operated within hierarchical systems, balancing central control with operational flexibility. Japanese feudal intelligence reflected similar organizational capabilities. The ninja families of Iga and Koga mastered specialized training, developed trade networks as covers, and preserved multi-generational traditions of espionage.

What stands out about Eastern intelligence organizations is their permanence. They weren't formed for specific wars and then disbanded afterward. They were permanent institutions evolving over centuries, allowing refinement of techniques and accumulated knowledge that Western ad hoc systems couldn't match.

Western intelligence structures remained relatively primitive until the early modern period. Renaissance Italian city-states established the first advanced Western operations, but even these lacked institutional depth.

It's a reflection of deeper cultural attitudes toward information. Eastern systems treated intelligence as a continuous necessity that required permanent institutions. Western systems often viewed it as a wartime requirement that could be improvised when needed and neglected during peace.

Connecting East and West

The Mongol Empire acted as a historical bridge between Eastern and Western intelligence traditions. Under Genghis Khan and his successors, Mongol operations blended Eastern philosophical comfort with deception and impressive organizational skills.

Their arrow riders weren't just spies—they were trained intelligence officers who could assess defense weaknesses, gauge political divisions among enemies, and track potential collaborators.

As Mongol conquests extended into Eastern Europe, Western powers experienced firsthand the devastating effectiveness of strategic intelligence. European kingdoms, used to gathering battlefield information through basic spying, suddenly faced an enemy that knew their movements, resources, and internal political divisions before armies even met.

This cross-fertilization accelerated during the Crusades, when European powers were confronted by advanced intelligence networks operated by Islamic states. The Assassins—the Nizari Ismaili sect—displayed the power of targeted operations and deep-cover agents to Western observers.

These cultural contacts contributed to the first major synthesis between Eastern and Western perspectives. European powers began incorporating Eastern methods, particularly by using merchant networks for intelligence and creating more permanent structures.

East Teaches West

European colonialism opened up new opportunities for Eastern intelligence techniques to influence Western practices. Colonial administrators in Asia frequently found themselves outmatched by local networks and were forced to adapt.

When the British East India Company set up in India, they ran into intelligence systems far more powerful than anything they had at home.

The Mughal Empire had elaborate spy networks, coded signals, and counterintelligence measures that impressed and alarmed European spies.

Similar encounters took place throughout Asia as European powers expanded their reach. In Japan, China, and Southeast Asia, colonial administrators became targets of operations that exploited Western tendencies toward direct approaches.

Colonial interactions led to significant changes. The British, in particular, developed new ideas as they absorbed Eastern practices. By the late 19th century, British operations had incorporated many Eastern elements. Officers adopted disguises, developed deep cover identities, and used psychological manipulation techniques.

This cultural exchange worked both ways. As Western powers extended their reach, Eastern states began incorporating Western influences, particularly in technological and organizational areas.

The Meiji Restoration in Japan marked a turning point. Japanese intelligence services consciously studied Western models while keeping their traditional strengths in human intelligence and deception. It was a powerful hybrid that surprised Western powers during 20th-century conflicts.

Technology Meets Tradition

The 20th century saw an accelerating convergence between Eastern and Western traditions. The Cold War, in particular, forced both blocs to adopt whatever methods proved effective, regardless of their cultural origins.

Soviet intelligence also integrated Eastern patience and deception with Western technology. The KGB's information warfare doctrine emphasizes stirring confusion and psychological disorientation, making targets distrust their own perceptions—we'll explore these tactics in more detail in Chapter 8. It's similar to Sun Tzu's emphasis on enemies defeating themselves through misperception.

Asian intelligence services remained true to their traditional strengths while incorporating Western elements. Chinese intelligence still thinks in decades rather than years. They're perfectly comfortable running operations that might not show results for a generation. Their long-term perspective stems from their historical roots, giving them an advantage that Western services, with their emphasis on quarterly results, struggle to match.

Reflections

The ancient Asian spies weren't playing checkers; they were playing 5D chess. Sun Tzu's spy categories read like a CIA manual written 2,500 years early. The ninjas were tracking targets while the West was still debating whether bathing was necessary. Chinese intelligence officers were running double agents before Rome was even a concept. These systems weren't primitive—they were carefully designed and highly effective. The Mongols created networks that would make today's intelligence agencies drool. Next time some tech company announces "revolutionary" secure messaging, remember: Japanese spies were using pocket-sized, hand-held throwing blades when Europeans thought the sun revolved around the Earth.

Chapter 4:

The Birth of Modern Intelligence

If you walked the streets of Florence in the early 1500s, you'd bump into Niccolò Machiavelli analyzing everything around him. This wasn't casual people-watching—he was tracking who spoke to whom, who was getting richer, and which families were making alliances through marriage. For him, gossip wasn't entertainment. It was currency.

Machiavelli knew something most politicians today still haven't figured out: Information beats money, armies, or fancy titles every time. He'd watch two diplomats smile and bow to each other in public while their spies were stealing each other's mail by night.

While artists were painting the ceiling of the Sistine Chapel, Machiavelli was writing what basically amounts to "Spying 101" for rulers. He saw princes and dukes crash and burn simply because they didn't know what was happening three cities over.

The Renaissance wasn't just about beautiful art and science breakthroughs. Those same curious minds were creating the first real intelligence networks. Looking over the next few pages, we'll see how Machiavelli's ideas about keeping tabs on friends and enemies alike created the blueprint for how countries still spy on each other.

Machiavelli's Influence on Intelligence Practices

You could say Machiavelli was Florence's original spymaster. This guy didn't just write about power from some ivory tower—he got his hands dirty, running actual spy networks across Italy. Think of him as the Renaissance version of George Smiley mixed with a political genius.

"The Prince" gets all the fame, but if you read between the lines, you'll find that Machiavelli essentially wrote the first spy handbook.

What made him so good? He'd been there, done that. As Florence's diplomat, Machiavelli traveled everywhere, meeting everyone from the Pope to backroom fixers. He built networks of informants in every major city, carefully sorting solid tips from useless gossip. He could tell when someone was feeding him false information just by how they presented it.

This wasn't abstract theory. Machiavelli would personally interview travelers from enemy territories, cross-check stories between different sources, and piece together what was really happening beyond Florence's walls.

Kings and queens across Europe devoured his ideas. They realized good intelligence could save their crowns (and sometimes their heads).

Picture Italy in the 1500s—there was madness everywhere, and backstabbing (literal and figurative) was the national sport. More than a philosophical writer, Machiavelli was Florence's top intelligence guy during the most cutthroat period in Italian history.

When sent as an ambassador to France or Rome, he'd be all smiles during official meetings, then spend his evenings in taverns buying drinks for local officials. After a few glasses of wine, these officials would spill secrets that changed Florence's entire foreign policy by morning.

He messed up a few times, too! One time, he trusted the wrong informant, and Florence almost went to war based on fake news. But what did Machiavelli do? He created the first known "source reliability rating system"—basically figuring out what went wrong, who was full of it, and who actually knew things.

His big innovation was turning every diplomat into a part-time spy. Before Machiavelli, diplomats just carried messages and looked pretty. He essentially said, "You're already there, might as well keep your eyes open!" He trained his diplomats to write two reports: the official one and the secret one with all the juicy stuff they'd overheard at parties or spotted while touring castles.

You know what kept Machiavelli up at night? Worrying about nosy people reading his mail.

Before his time, sending secret information was insanely risky. A messenger carrying important news could get robbed, bribed, or just plain curious. Your national secrets are now public knowledge. Machiavelli turned this whole game upside down.

He became obsessed with creating foolproof ways to hide messages. His agents used everything from simple letter substitutions to wild chemical concoctions. One technique involved hiding real messages within boring, routine letters about trade or family matters. The actual intelligence was tucked between lines of innocent small talk.

What's amazing is how his ideas spread like wildfire across Europe. Kings who publicly criticized Machiavelli were privately stealing his encryption methods! You can draw a direct line from Machiavelli's hidden ink formulas to modern encryption. Those little lock symbols on your banking apps? That "end-to-end encryption" on your messages? They're just fancier versions of what this Renaissance political genius figured out with ink and paper 500 years ago.

Evolution of Formal Intelligence Services Under Elizabeth I

Machiavelli was legitimately concerned about secret messages being intercepted, which was a real problem in Renaissance politics. He advocated for encryption methods that were advanced for the 1500s. His diplomatic service for Florence included creating secure communication channels using period-appropriate techniques like coded language and trusted courier networks.

Francis Walsingham came along in Elizabethan England and transformed intelligence gathering into a systematic operation. As Queen Elizabeth's Principal Secretary, he established what historians consider the first professional intelligence service in England.

Walsingham maintained a network of paid informants across Europe who reported on threats to England, particularly from Spain and supporters of Mary, Queen of Scots. His agents truly did infiltrate foreign courts, intercept communications, and track suspicious Catholics during a time of intense religious conflict.

His most famous success was uncovering the Babington Plot in 1586. Walsingham's agents intercepted encoded letters between Anthony Babington and Mary, Queen of Scots, revealing a conspiracy to assassinate Elizabeth. His team deciphered these messages and, with Elizabeth's approval, allowed the plot to develop until they had sufficient evidence to arrest everyone involved.

Walsingham essentially invented the spy game as we know it. He set up something revolutionary. Eyes and ears in all the right places.

This guy placed spies inside foreign embassies right in London. They'd pose as servants, copying documents when ambassadors weren't looking or listening at doors during important meetings. Pretty gutsy for the 1570s.

But Walsingham went further. He sent English merchants abroad with secret second jobs. They'd sell wool in Spanish ports while quietly counting warships or tracking troop movements. Some of his agents could perfectly fake being devout Catholics, infiltrating groups plotting against Protestant England.

His favorite trick was using double agents. He'd catch someone working for Spain, then flip them by offering money or pardons. These people would feed false info back to Spain while reporting everything to Walsingham. The Spanish thought they had spies in England, but really, Walsingham controlled what information they received.

What made Walsingham special was being proactive. Rather than waiting for attacks, he hunted for threats before they materialized. His motto might as well have been "know your enemy before they strike."

His methods became standard practice. The cipher systems, agent networks, and intelligence analysis techniques he developed became the foundation for how governments collected and used secret information for centuries afterward.

Strange to think—many basic principles of modern intelligence work trace back to this bearded Elizabethan secretary.

Even with this brilliant start, spying wasn't perfect. Far from it. Intelligence systems across Europe faced some embarrassing failures. Countries that thought they'd copied Elizabeth's spy methods quickly learned they hadn't quite gotten it right.

Impact of Intelligence Failures on War Outcomes

Before Walsingham, kings basically showed up to battle and hoped for the best. After him? Warfare got sneakier, smarter, and much more calculated.

The Thirty Years' War (1618–1648) really shows this shift. Cardinal Richelieu (France's version of Walsingham, but with fancier hats) created what he called the "Cabinet Noir"—literally the "Black Cabinet." This wasn't furniture—it was a secret team that steamed open letters, broke codes, and copied diplomatic messages.

Richelieu's spies were everywhere. They infiltrated enemy camps, bribed servants in Spanish war councils, and created networks across Europe. With this advantage, France could make alliances at exactly the right moments and attack when enemies were weakest.

Military commanders started caring about a whole new set of questions. What's the enemy's food supply like? Are their soldiers getting paid? Are there disagreements between their generals? This kind of intel often proved more valuable than knowing how many cannons they had.

By the late 1600s, every major army had intelligence officers attached to field commanders. Being a good general wasn't just about courage anymore—it was about having better information than your opponent.

Walsingham would've been proud. His spy experiments had grown into sophisticated systems that literally decided who won or lost wars.

The Sack of Magdeburg

The Sack of Magdeburg in 1631 shows what happens when spy networks fail. It's one of history's most brutal examples of "you snooze, you lose" in warfare.

Magdeburg was a Protestant German city that thought it was safe. Their leaders received reports that Catholic imperial forces were busy elsewhere. Wrong! The intelligence was completely off. While Protestant commanders read outdated reports, the Catholic army had already surrounded the city, led by Johann Tserclaes, Count of Tilly—a veteran Flemish commander who had crushed Protestant forces in numerous battles during the Thirty Years' War.

The consequences were catastrophic. When Tilly's forces broke through, they burned Magdeburg to the ground and killed thousands of civilians—nearly the entire population. Survivors said soldiers went from house to house, slaughtering families and setting fires. The massacre shocked Europe.

What went wrong? The Protestant intelligence system relied on travelers and merchants to spread news. Too slow! They had no dedicated spies watching enemy movements. And the few warnings they did receive were dismissed as exaggerations.

Albrecht von Wallenstein

Albrecht von Wallenstein is proof of how effective intelligence was used during this time. As Imperial commander, Wallenstein developed an extensive information network that served him throughout his military campaigns. His intelligence system included paid informants in various cities and strategic locations.

Wallenstein hired traders, travelers, and local officials who could move freely and gather information without suspicion. His network allowed him to make informed military decisions based on enemy troop movements and supply situations.

The efficiency of his intelligence operations gave him significant advantages in positioning his forces. For example, historical records show he successfully predicted enemy movements on multiple occasions. He could then choose favorable terrain for encounters or avoid unfavorable confrontations altogether.

The contrast between Magdeburg's fate and Wallenstein's successes showed military leaders of the era that reliable intelligence could mean the difference between victory and disaster.

The Evolution of Military Intelligence

Those early intelligence lessons contributed to how wars are fought today.

After the Magdeburg disaster, armies created actual intelligence departments with defined reporting lines. They stopped relying on random travelers for news and began using dedicated agents with specific responsibilities.

These basic lessons were built over time. By the American Civil War, both sides had signal corps that intercepted telegraph messages. During World War I, radio intercepts became standard practice. These developments were not accidental. They were a direct result of lessons learned from past intelligence failures.

A perfect example is how the Allies used intelligence during the liberation of Paris in 1944. The intelligence work wasn't just about having spies on the ground (though they had those, too). The Allies had learned from centuries of experience that you need multiple sources of information.

They combined French Resistance reports with aerial photography, prisoner interrogations, and, most importantly, intercepted German communications. Thanks to the Ultra program that broke German codes, Allied commanders knew exactly where German tanks were positioned and which roads were open.

General Patton once said, "The object of war is not to die for your country but to make the other bastard die for his" (Patton, n.d.). After everything that happened in Magdeburg, those hard-earned lessons had finally sunk in.

Reflections

The spy game we play today was born in the smoky back rooms of Renaissance Florence.

Before Machiavelli and his sneaky friends came along, kings basically went to war because someone insulted their hat. After these Renaissance masterminds? Rulers started demanding actual facts before making decisions. Revolutionary!

You have to love the parallels between then and now. Those Italian spy networks from the 1500s were essentially CIA stations without the high-tech equipment. They recruited local sources, paid for information, and sent coded messages back home. Sound familiar?

The best part is how some tricks never get old. Renaissance spies would pose as traveling businessmen or scholars to slip across borders unnoticed. Fast forward 500 years, and intelligence agencies still use "commercial cover" to place officers in foreign countries. The tools change, but the tricks stay the same.

What's truly mind-blowing is how these Renaissance innovations directly saved lives throughout history. During the Cold War, intelligence agencies on both sides used techniques that Walsingham would have instantly recognized. The CIA's recruitment of Soviet officials? Pure Renaissance playbook. The KGB's use of dead drops and coded messages? Straight from the 1500s.

These historical connections aren't just cool trivia. They're survival lessons. Every major intelligence failure in modern times (such as Pearl Harbor or 9/11) happened when countries forgot the basic principles learned during the Renaissance: verify information from multiple sources, don't just hear what you want to hear, and always build networks before you need them.

If you read about some international intelligence scandal, it's not just politics. It's a game that's been played since Machiavelli taught rulers that knowledge is power.

Chapter 5:

Revolutionary Secrets—

Intelligence in the Age of Upheaval

History books celebrate the roar of cannons at Yorktown and the cries of Parisian mobs. However, they seem to forget the conversations that actually shaped nations.

For every famous revolutionary speech, a dozen coded messages changed fate more decisively. Secret agents—shopkeepers, housewives, and servants with extraordinary nerve—altered our world more significantly than the generals who claimed victory medals.

What if a single intercepted letter had collapsed the American Revolution? What if just one British spy had discovered Washington's crossing before Trenton?

This chapter reveals the incognito fighters. We track Benjamin Tallmadge as he built America's first organized intelligence network from scratch. We follow the paranoid vision of Robespierre's Committee of Public Safety, which created surveillance systems so thorough that they monitored what citizens discussed at dinner tables. We watch these makeshift spy rings evolve into powerful machines that would define national security for centuries.

Benedict Arnold's story transforms from simple treachery into something far more valuable. A case study in the fatal flaws that still plague intelligence operations today.

Black spies like James Lafayette and Harriet Tubman risked everything for a country that denied them basic rights. These forgotten heroes represent America's great paradox: Those who had the least freedom often risked the most to secure liberty for others.

Most fascinating are the drawing rooms of Vienna in 1815, where diplomats smiled through gritted teeth while spies stole secrets that forever changed Europe. Metternich's group of informants—from servants to aristocratic mistresses—gave Austria advantages worth more than armies.

It becomes obvious that the strongest victories were won by forgotten patriots who understood that information, properly applied, could overthrow kings more effectively than any guillotine.

Intelligence Foundations in Early Revolutions

The guns of Yorktown tell half the story. The other half hides in plain sight.

Ever wonder how a ragtag colonial army beat the world's mightiest empire? Sure, French help mattered. Valley Forge showed grit. But something else tipped the scales—something history books barely mention.

Washington bet America's future on people who never fired a shot. "Upon secrecy, success depends in most enterprises," he wrote, sometimes spending more on spies than bullets (Washington, 1777). This Virginia general who lost more battles than he won somehow stayed one step ahead of British commanders who outgunned him at every turn.

His secret weapon was a 26-year-old Yale graduate named Benjamin Tallmadge and a network of everyday people doing extraordinary things.

Tallmadge built America's first real spy ring from friends he grew up with on Long Island. His agents looked like everyone else. Abraham Woodhull sold vegetables to British officers from his farm wagon. Austin Roe made deliveries that took him through British checkpoints with messages hidden in his saddle.

What they created was genius in its simplicity. They wrote letters in invisible ink that appeared only when treated with a special chemical. They used coded messages where "355" meant their female operative who charmed her way through British social circles, collecting careless comments at dinner parties (Bleyer, 2022).

Anna Strong, a farmer's wife, hung her laundry in ways that signaled to agents when couriers had arrived—certain items signaled specific meeting spots. British troops walked past daily, seeing a housewife doing chores.

These seemingly ordinary Americans changed history. They exposed Benedict Arnold's plot to surrender West Point three weeks before he could pull it off. They uncovered British plans to attack French allies. They exposed a British scheme to flood America with counterfeit money and crash the economy.

The most amazing part? Almost none were caught. British spy networks collapsed regularly with agents dangling from gallows. Meanwhile, Tallmadge's network operated right under enemy noses for years. Many took their role in winning independence to their graves. Their contributions were only discovered when historians decoded Washington's papers decades later.

The genius of Tallmadge's spies was their everyday cleverness. They created spy gear from household items that British soldiers ignored.

Take their invisible ink. Unlike regular ink, this special formula vanished completely when dry. A letter about crop prices or family gossip could pass through British checkpoints, looking perfectly innocent. But that same letter, when brushed with a chemical mixture, suddenly revealed military plans.

Washington loved this trick. One spy would write a normal letter with visible ink. Later, another agent filled the blank spaces with secret messages using the invisible solution. British officers who intercepted these letters saw nothing suspicious. To them, it was just boring farmer talk.

The recipe for this invisible ink stayed secret for 200 years. We now know James Jay (brother of founding father John Jay) created it using gallic acid from oak apples. The revealing agent was a simple iron sulfate solution that turned the hidden writing bluish-black on contact.

Coded messages added another layer of protection. Tallmadge created a system where numbers replaced names and locations. Washington was always "711." New York became "727." British troop movements were disguised as business transactions. A message about "purchasing 712 items from 728" actually meant British ships were heading to Rhode Island (*The Culper Code Book*, n.d.).

Robert Townsend (code name: Culper Jr.) perfected these methods. Working as a merchant by day, he collected information from loose-lipped British officers drinking in his family's business. At night, he wrote secret messages using number codes and invisible ink, then hid them inside blank pages of financial ledgers.

For extra security, the spies created their own dictionary. Common words meant something completely different to Culper agents. "Stocking merchant" meant military supplier. "Factory goods" meant military intelligence.

These primitive but effective tools worked because they matched the technology of the time. British counterintelligence looked for physical evidence. Hidden papers or suspicious behavior. They never imagined that the boring letter about cabbage prices they approved actually included plans for Washington's next attack.

The Group Behind the Revolution's Secret War

Fancy symbols and invisible ink captured imagination, but Tallmadge's true genius was people management. He built a network where farmers trusted merchants who trusted sailors who trusted housewives with their lives. Each person knew just enough to do their job but not enough to pose a threat to each other.

The stakes? A slow death by hanging. Just ask Nathan Hale, executed by the British after his failed spy mission with his famous last words: "I only regret that I have but one life to lose for my country" (*Rev. War Biography: Nathan Hale*, 2019).

Tallmadge never forgot that lesson. He recruited childhood friends with whom he had deep personal bonds. Woodhull and Townsend had known each other since boyhood. Their trust was built over decades, not through patriotic speeches. Strong was connected to Woodhull by marriage. These connections created loyalty that British sterling couldn't buy.

The Ring's structure was brilliantly compartmentalized. Woodhull gathered intelligence but never wrote reports—too dangerous with British soldiers boarding at his house. Instead, he memorized information and passed it to Austin Roe during fake business meetings. Roe carried nothing suspicious while riding through checkpoints. Only when it was safe would he pass intelligence to Townsend, who encoded everything.

British counterintelligence was everywhere. They hanged suspected spies publicly, leaving bodies displayed as warnings. They planted double agents and paid informants for leads. One slip meant death.

Tallmadge countered by creating false identities for his agents. Woodhull posed as a Loyalist sympathizer, openly complaining about rebels. Having this cover let him move freely through British-controlled areas. Townsend's business catered to British officers, which made him appear loyal and evade the enemy's suspicion.

The payoff came in military advantage. When the Culper Ring discovered British plans to ambush newly-arrived French forces at Rhode Island, Washington reinforced the area before the British arrived. When they learned of Clinton's planned attack on West Point through Benedict Arnold's treachery, they stopped it before it began.

One event remains the most telling of the network's success: After the war, British intelligence officials expressed shock at learning the identities of Culper agents. They knew they were surrounded by spies, but they never suspected these particular individuals.

The greatest tribute to Tallmadge's human network is that not one Culper Ring member was ever caught. In a war where spies regularly swung from gallows, this flawless record came from something beyond clever gadgets—it came from picking the right people and giving them just enough knowledge to succeed without enough to betray others under torture.

Triumphs and Betrayals

The secret war gave Americans something bullets couldn't: confidence. Each hacked British message proved the rebels could outsmart the world's greatest empire. With every battle avoided through advance warning and every trap escaped through leaked plans, colonial leaders realized they weren't just lucky—they were getting good at this.

These weren't just tricks by outmatched colonists. They were creating something new: an American approach to intelligence. Washington's agents used local knowledge and tight community bonds to outmaneuver British spies who stuck out in colonial towns. The British had military might, but Americans had home-field advantage and used it brilliantly.

Yet for all their successes, the war had a dark side. The Benedict Arnold case exposed dangerous blind spots.

Arnold—hero of Saratoga and loyal commander—nearly handed West Point to the British in 1780. His betrayal blindsided Washington and showed how personal grudges could undermine intelligence work. A decorated general feeling overlooked and underappreciated proved more dangerous than any British spy.

The British played this game well. Major John André, their intelligence chief, studied Arnold for months, noticing his complaints about Congress, financial troubles, and his burning ambition. They offered him money and rank in exchange for America's most important fortress.

What makes this story fascinating is how close it came to working. Arnold's plot wasn't discovered through brilliant counterintelligence work but by pure accident. Three militiamen stopped André on a road and found plans hidden in his boot.

This near-disaster taught Washington painful lessons. Having good spies watching the enemy wasn't enough—you needed to watch your own side, too. After Arnold, Washington added counterintelligence to his operations, checking backgrounds more carefully and limiting access to sensitive information.

The story of America's Revolutionary spies shows both incredible innovation and sobering reality. The same network that brilliantly gathered enemy secrets missed the traitor in their ranks. They created techniques that would serve intelligence agencies for centuries but couldn't prevent human nature from nearly sabotaging everything.

America's Invisible Patriots

While we celebrate generals who won our wars, America's most daring heroes worked in the dark, risking torture and death without recognition, medals, or glory.

Meet James Lafayette, an enslaved man who slipped through British lines during the Revolution, gathering intelligence that helped Washington win at Yorktown. Then there's Harriet Tubman—you know her as the Underground Railroad conductor, but did you know she ran spy networks that penetrated deep into Confederate territory?

Being Black in America made them perfect spies. The same society that considered them invisible gave them the perfect cover to move undetected. White officers spoke freely around "just another slave." Confederate soldiers never suspected the unassuming Black woman was memorizing troop numbers.

America's freedom was secured by those who had none themselves.

The Spymaster Who Spied America to Freedom

"The Negro is providing us with the best intelligence we have," wrote the Marquis de Lafayette to George Washington in the summer of 1781. The "Negro" he referred to was James Lafayette—a Virginia slave who had infiltrated British headquarters and was secretly transmitting information that would soon bring the American Revolution to its victorious conclusion (Gruber, 2022).

History rarely remembers its greatest spies. This statement is doubly true for Lafayette, whose extraordinary service remained largely hidden for two centuries—buried by racial prejudice and the inherent secrecy of espionage. Yet few individuals did more to secure American independence.

In the war's desperate final act, British General Cornwallis had established headquarters at Yorktown, Virginia. Among the local enslaved people seeking work in his camp was a man who appeared eager to escape his American masters. The British officers barely noticed him—just another Black face among the servants polishing boots and emptying chamber pots.

This invisibility was Lafayette's greatest weapon. As he moved silently through headquarters, officers discussed troop positions, fortification weaknesses, and naval signals with complete openness. Each detail lodged in Lafayette's remarkable memory. Under the cover of hunting trips, he regularly slipped away to pass this intelligence to American couriers waiting in the Virginia woods.

He risked far more than most patriots. Discovery meant not just execution as a spy but likely torture for a Black man caught betraying the British, who had promised freedom to slaves who joined their side.

Lafayette's intelligence proved decisive when Washington's forces attacked Yorktown in September 1781. American artillery targeted the areas where the British defenses were weakest—positions identified in Lafayette's reports. When Cornwallis surrendered his entire army on October 19th, few knew that an enslaved man had helped engineer this stunning victory.

The bitter irony? After risking everything for American liberty, Lafayette returned to slavery. His freedom came only in 1786, when the Marquis de Lafayette personally intervened with the Virginia legislature, testifying to his spy's essential services and perfect fidelity. Granted his liberty at last, he adopted the surname "Lafayette" in gratitude to his French patron.

In 1824, during Lafayette's triumphant return tour of America, the elderly James visited his former commander at a Richmond reception. Witnesses described the emotional moment when the celebrated French general, spotting his former spy across the crowded room, rushed to embrace him before Virginia's stunned elite. A rare public acknowledgment of a Black patriot's significant contribution to American independence.

James Lafayette's story reminds us that American freedom was secured through the courage of those who were denied that freedom. By turning the very prejudices of colonial society into his strategic advantage, this forgotten hero helped birth a nation whose founding principles—however imperfectly realized—would eventually extend liberty to all.

Civil War Spy, Scout, and Military Strategist

The Confederate officer barely glanced at the old Black woman shuffling along the roadside outside his fort. Her face half-hidden beneath a tattered bonnet, she was just another "auntie" selling eggs to supplement her meager existence. He would have been shocked to learn she was memorizing every detail of his defenses—counting cannons, keeping track of guard rotations, and identifying blind spots in the defensive perimeter.

She wasn't just any spy. She was Harriet Tubman.

While history remembers Tubman's heroic midnight journeys on the Underground Railroad, guiding approximately 70 enslaved people to freedom, her remarkable career as a Union intelligence operative has remained largely in the dark. Yet during the Civil War, this illiterate former slave ran one of the most effective spy networks of the conflict. She orchestrated a devastating military raid, becoming the first woman in American history to plan and lead a military operation.

In early 1862, Tubman arrived in the Sea Islands region at the request of Massachusetts Governor John Andrew, who recognized her potential value to Union forces struggling to gain traction in the Confederate heartland. What military commanders encountered was something far beyond their expectations: a tactical genius with unparalleled knowledge of covert movement.

Tubman built her spy ring from people the Confederates deemed invisible—formerly enslaved men and women who could move unnoticed through Southern territory. Her agents posed as field workers, laundresses, and cooks, exploiting the blind spot in Confederate security—the assumption that Black people were too simple to gather military intelligence.

Her crowning achievement came on June 2nd, 1863, along South Carolina's Combahee River—a raid so brilliantly executed that it represents one of the most successful intelligence and combat operations of the war.

Working closely with Colonel James Montgomery of the Second South Carolina Volunteers, Tubman plotted an assault that would strike deep into Confederate territory. In the pre-dawn darkness, she guided three Union gunboats upriver, personally indicating where torpedoes lurked below. As the gunboats approached the plantations, Tubman stood on deck and began singing spirituals—coded messages that signaled to hiding slaves that liberation had arrived. The response was electric.

By operation's end, 756 enslaved people had been liberated—many of the men immediately enlisting in the Union Army. Five plantations lay in ashes, along with large stores of food and cotton bound for Confederate forces. Confederate troops arrived too late to do anything but witness the devastation of what newspapers dubbed "the Combahee Ferry Raid."

She received just $200 for years of dangerous service—less than many male scouts earned in a few months (Maranzani, 2013). Though General Rufus Saxton praised her invaluable service to the Union cause, the War Department repeatedly denied her claims for a military pension. It took until 1899—34 years after the war ended—for Congress to finally grant her $20 a month, and even then, only as the widow of a Union veteran she had married after the war.

What makes Tubman's intelligence operations even more remarkable were the obstacles she overcame. Born into slavery, unable to read or write, and suffering from narcolepsy caused by a childhood head injury (an overseer fractured her skull), she relied on extraordinary memory, quick thinking, and uncanny intuition to outfox Confederate forces.

Espionage Strategies During the French Revolution

Benedict Arnold's betrayal showed America the dangers of a single traitor. But what happens when an entire government turns paranoia into policy? The French Revolution gives us that chilling answer.

After overthrowing their king in 1789, the French faced enemies on all sides—foreign armies at their borders and suspected traitors hiding among citizens. Their response created something far different from Washington's careful intelligence gathering: a surveillance state where accusations became weapons and suspicion meant death.

The Committee of Public Safety, led by Robespierre, transformed intelligence from a military tool into an instrument of control. While Washington's spies watched British troops, French revolutionary spies watched French citizens. The Committee recruited thousands of informants to report on neighbors, family members, and colleagues who seemed less devoted to revolutionary ideals.

These weren't trained operatives but ordinary people—grocers keeping track of which customers complained about bread prices, servants reporting conversations overheard in wealthy homes, children encouraged to expose parents who said the wrong things. One careless comment about "the good old days" might bring armed men to your door at midnight.

The system worked with ruthless efficiency. Thousands were arrested based on these civilian reports. The infamous Law of Suspects made "showing concern about the Revolution" enough for imprisonment. It wasn't targeted intelligence but mass surveillance designed to terrify the population into compliance.

But there was a tragic flaw. Much of this "intelligence" was garbage. People settled personal grudges by denouncing rivals. Jealous neighbors reported the family with slightly better food. Workers attacked bosses they disliked. These reports led to the Reign of Terror, in which thousands of accused "enemies of the Revolution" met the guillotine without any evidence beyond accusations.

Meanwhile, actual threats went unaddressed. British agents freely funneled money to royalist groups in western France, creating a civil war that nearly ended the Republic. When France sent armies against European powers, they marched with outdated maps and inaccurate information about enemy positions, leading to devastating defeats in 1792 and 1793.

The Committee confused quantity of information with quality. Their filing cabinets overflowed with reports about citizens' political comments, while they missed critical military intelligence. When the Revolutionary Army lost battles due to surprise attacks, leaders blamed "traitors" rather than their own intelligence failures.

This dark chapter shows how thin the line can be between intelligence and tyranny. Where Washington used spies to fight an external enemy, the French used them to terrorize their own people. Their system gathered enormous amounts of information but failed at what intelligence is supposed to do: separate valuable facts from noise and prejudice.

Shifting Loyalties: The Spy Games of Revolutionary France

General Charles François Dumouriez embodied the chaotic loyalties of revolutionary France. A talented military commander who won important early victories for the Republic, he later fled to the Austrian forces with plans to march on Paris and restore the monarchy. His dramatic defection revealed how personal conviction could trump national loyalty in revolutionary chaos.

"I served France, not her butchers," Dumouriez most likely declared when explaining his betrayal. Having watched the revolution devour its children—with moderate voices silenced by the guillotine—he concluded the Committee of Public Safety had become a greater threat than foreign armies.

This constant shifting of allegiances created a spy's paradise. People changed sides not only because of bribery or threats but out of genuine conviction as the revolution's ideals gave way to bloodshed. One week's patriot became next week's traitor as political winds shifted.

Revolutionary France became a tangle of secret networks. Catholic priests operated underground chains to smuggle aristocrats to safety. British gold funded royalist rebellions in western provinces. Austrian and Prussian agents collected military intelligence from disillusioned officers considering Dumouriez's path.

The revolutionaries fought back with creative tactics. They planted false information in royalist circles, watching who acted on it to identify groups. They created counterfeit British currency to trace how enemy funds flowed into France. They seized mail and developed simple but effective code-breaking techniques to decipher royalist messages.

What makes this period fascinating is how personal and political betrayals intertwined. When the powerful revolutionary Georges Danton was executed, it wasn't just for alleged treason but because rivals like Robespierre feared his influence. During internal power struggles, intelligence was used as a weapon to destroy political enemies and also to catch actual spies.

This ruthless baptism transformed French intelligence from amateur enthusiasm into professional expertise. By the time Napoleon seized power in 1799, he inherited battle-tested methods. Where revolutionary spies had worked with passionate deception, Napoleon built systematic networks with clear chains of command and professional standards.

His Bureau of Statistics—actually a foreign intelligence service despite its bland name—deployed agents across Europe disguised as traders and diplomats. Napoleon's battlefield intelligence proved revolutionary; while earlier generals relied on scouts reporting what they personally observed, Napoleon created intelligence cells that compiled reports from multiple sources to create detailed pictures of enemy positions and plans.

Most importantly, Napoleon mastered disinformation. Before invading Egypt, he leaked false plans suggesting his fleet was headed for

England. Before Austerlitz, he deliberately presented his army as weak and demoralized, enticing enemies into a trap.

Expansion of Intelligence in Major Historical Conflicts

While French blood still stained Paris streets, a quieter battle began across Europe. The British—facing Napoleon's ambition to dominate the continent—weren't fighting with just ships and soldiers but with words and codes.

The British Admiralty built something new: a permanent intelligence service run by professionals instead of enthusiastic amateurs. Their system had three major advantages. First, they placed agents in every major European port who tracked French naval movements, letting British ships intercept French fleets before they combined forces. Second, they captured diplomatic messages between France and its allies, breaking codes to reveal Napoleon's true plans. And third, they planted false information that sent French forces on wild goose chases.

This was a coordinated system. The Alien Office tracked foreign visitors to Britain, spotting French agents through careful observation. The new telegraph network rushed intelligence from coastal lookouts to London within hours instead of days, giving military planners precious time to act.

Captain Sir George Scovell represented this new skill perfectly. A language prodigy, he broke the "Great Paris Cipher"—Napoleon's most complex military code—allowing Wellington to read the emperor's orders to his generals in Spain almost as soon as they did. This information edge helped smaller British forces defeat larger French armies again and again.

Meanwhile, at the 1814–1815 Congress of Vienna, where Europe's leaders gathered to rebuild the continent after Napoleon's defeat, a different kind of spy game played out in glitzy ballrooms.

Charles Maurice de Talleyrand, representing defeated France, turned weakness into strength through brilliant social espionage. He hosted lavish dinners where wine flowed freely, creating settings where diplomats spoke carelessly. His network of charming women— strategically placed in social circles—collected bedroom talk from visiting officials. But his real skill was letting others hang themselves through unguarded words. Though officially shut out of key meetings as a representative of a defeated nation, his intelligence network kept him better informed than those who attended.

The results were phenomenal. Though France lost everything on the battlefield, Talleyrand's backroom maneuvering saved its territory and standing. By cleverly exploiting tensions between Britain, Russia, Austria, and Prussia—tensions he learned about through his social spy network—he stopped France from being carved up and kept it among Europe's powers.

These changes permanently shifted how nations viewed intelligence. The British showed that ongoing counter-intelligence work could protect a nation better than armies alone. Talleyrand proved that in diplomatic fights, knowing others' secrets is more effective than having the strongest position.

These innovations turned spying from a wartime necessity into a permanent government function.

Reflections

They left no monuments. No medals glisten on their chests in grand portraits. The most successful among them died with neighbors thinking they'd led unremarkable lives.

Yet these invisible hands—America's first spies, France's ruthless informants, Europe's diplomatic eavesdroppers—shifted revolutions more decisively than many generals.

Washington's greatest weapon wasn't his army but his spy network. When British commanders moved troops by the thousands, Washington knew before the dust settled on colonial roads. He knew because a nervous shopkeeper named Townsend paid attention while serving drinks to British officers. He knew because a farmwife named Strong hung laundry in a way that signaled "message ready for pickup." The revolution succeeded partly because farmers fought, but largely because farmers spied.

The French took different lessons from America's success. Where Washington built networks based on trust, the revolutionary Committee created systems built on fear. Their approach worked, but at a terrible cost. The problem wasn't that they used surveillance—it was that they collected so much information, they couldn't tell real threats from petty complaints. They showed how easily the shield of intelligence becomes the sword of terror.

Napoleon learned from both examples. He created the first modern intelligence service with well-defined command chains and professional standards. His Bureau of Statistics—actually a spy agency with a boring name—placed agents across Europe. Before battles, he spread false information so cleverly that enemies marched straight into his traps. The British countered by breaking his codes and reading his orders sometimes before his own generals did.

Then came Talleyrand, perhaps the greatest spy master of all. Officially just a diplomat representing defeated France, he rebuilt his country's standing through dinner parties. While others negotiated with formal proposals, he collected pillow talk from a circle of charismatic women who warmed the beds of Europe's decision-makers. His intelligence coup at the Congress of Vienna saved France from being dismembered after Napoleon's fall.

What binds these stories together? The recognition that information—who has it, who doesn't—determines winners and losers more surely than guns or gold. The colonial agent who learned of British troop movements saved more American lives than any battlefield hero. The French informer who misidentified neighbors as "enemies of the revolution" killed more innocent people than many soldiers.

Every encrypted message, every anonymous source, every classified document today follows guidelines set by these early practitioners. They showed that true power often lies not with those making the loudest noise, but with those collecting the quietest whispers.

Chapter 6:

Intelligence in the Age of Total

War

In 1943, Allied forces sank 51 German U-boats in the Atlantic—a devastating blow to Nazi Germany's naval strategy. Unbelievably, a large number of these victories came directly from intelligence gathered by codebreakers at Bletchley Park. A collection of 10,000 unlikely spies—mathematicians, linguists, chess champions, and puzzle enthusiasts—had altered world history through brainpower alone. Their existence remained classified for 30 years after the war ended, making it perhaps the most effective secret operation in history.

This was a new kind of warfare. The German captains never saw their real opponents—not the destroyers that depth-charged their submarines, but the brilliant minds who tracked their movements by breaking supposedly unbreakable codes. Intelligence used to support military operations, but now it was becoming the deciding factor.

World War II transformed espionage from a limited function into a central pillar of military strategy. Wars would now be won not just on battlefields, but in converted mansions where academics hunched over encrypted messages. Victory also came from hidden organizations running double agents and the courageous efforts of operatives—many of whom were women—who risked everything behind enemy lines.

The reality of "total war" created the need for "total intelligence"— comprehensive systems that collected, analyzed, and weaponized information at an unprecedented scale. Traditional spying merged with new technologies to create something entirely different. Radio direction finding pinpointed enemy transmitters. Aerial photography revealed hidden facilities. Computing machines tackled codes too complex for human minds alone.

This shift brought intelligence from the edges to center stage. When Allied forces landed at Normandy, their confidence came not just from superior numbers but from superior knowledge. They knew where German divisions were positioned. They knew which beaches were most heavily defended. They even knew which German commanders were away from their posts for a war game exercise on June 6th. What happened wasn't the result of chance—it was a major intelligence effort that had penetrated German communications on every front.

Intelligence changed warfare forever. From Mata Hari's controversial execution in 1917 to Turing's codebreaking brilliance in WWII, espionage and cryptography became strategic weapons like no other.

Espionage in World War I

The firing squad took shots at a woman who refused a blindfold. She blew a kiss to her executioners before their bullets found her heart. The date was October 15th, 1917, and with Mata Hari's death, one of history's most enduring spy myths was born.

Everything about her execution made for perfect propaganda. She was a Dutch exotic dancer with lovers from every European army, convicted of espionage against France.

What better symbol of the hidden enemy? Newspapers painted her as a diabolical seductress who used her body to extract military secrets, causing the deaths of thousands. Her name became synonymous with the femme fatale spy. Her legacy continues through countless books and films.

The truth? She was barely a spy at all.

Recently declassified intelligence files reveal a far less dramatic reality. After her dancing career faded, Mata Hari (born Margaretha Zelle) accepted money from the Germans to report gossip from her many admirers in French military circles. The intelligence she passed was trivial, mostly information anybody could read in newspapers. She had no cipher, no spy craft, and no access to anything important.

French intelligence knew this. Hari's conviction relied on deciphered messages that historians consider highly questionable, possibly even fabricated. But France in 1917 desperately needed a public victory and a scapegoat for military failures. Hari—foreign, sexually liberated, and with German connections—fit perfectly. She was worth more to them dead than alive. Her execution boosted French morale while warning other potential informants. The mythmaking began immediately, transforming an amateur informant into history's quintessential spy temptress.

Meanwhile, across the Channel, a very different kind of intelligence operation was changing history without glamor or headlines.

Room 40: The Real Game-Changer

In a cramped, hot room at the British Admiralty, a peculiar collection of people—crossword enthusiasts, language professors, and chess champions—hunched over German naval messages. They called themselves Room 40.

Unlike Hari's amateur efforts, Room 40 represented the future of intelligence—systematic, technical, and devastatingly effective. Their breakthrough came early in the war when the Russians recovered codebooks from a grounded German cruiser and shared them with Britain. Soon, Room 40 was reading Germany's naval communications almost in real time.

The results were immediate and dramatic. When German battle cruisers planned surprise bombardments of British coastal towns, Room 40's warnings allowed the Royal Navy to stop them. When the German High Seas Fleet moved, Room 40 knew before they left the harbor.

Their crowning achievement came in 1917 with the Zimmermann Telegram—a coded message proposing an alliance between Germany and Mexico against the United States. Room 40 was able to decrypt it without revealing their capabilities. When President Wilson saw the telegram's contents, it became inevitable that America would enter the war.

Room 40 achieved what historians consider the greatest intelligence triumph of its era. It fundamentally changed the course of the Great War and quite possibly saved Britain from defeat against German naval power. Their approach to codebreaking and signals intelligence would later inspire and inform the even more sophisticated Ultra operation that was decisive two decades later in the fight against Nazi Germany.

The contrast between Mata Hari and Room 40 reveals how World War I transformed the field of espionage. The old model that involved individual agents using personal charm to gather information gave way to technical intelligence using mathematics, technology, and organization. Individual spies still mattered, but the future belonged to decoders, radio interceptors, and analysts.

Yet while Britain pioneered signals intelligence, Germany was perfecting a different method to covert operations—one focused not on gathering information but on direct sabotage of enemy resources.

The Black Tom Disaster

At 2:08 a.m. on July 30th, 1916, residents across New York City woke up to what many thought was an earthquake. Windows shattered 25 miles away. The Statue of Liberty took shrapnel damage that remains visible today. The blast—equal to an earthquake measuring 5.5 on the Richter scale—wasn't natural. It was sabotage.

German agents had just blown up the Black Tom munitions depot in New Jersey, destroying two million pounds of ammunition bound for Britain and Allied forces. The explosion killed seven people, injured hundreds more, and caused $20 million in damages (equivalent to over $500 million today). It was the most destructive act of foreign terrorism on American soil until 9/11.

This wasn't a lone incident. German saboteurs targeted American factories, ships, and military installations in a coordinated campaign before the U.S. officially entered the war. They planted time bombs on cargo ships. They spread anthrax among horses being shipped to Allied forces. They even tried to blow up the international bridge at Vanceboro, Maine.

Yet these operations reveal more about the challenges of running foreign agents than their successes might suggest. The Black Tom explosion, while dramatic, actually accelerated America's path to war. Security measures were tightened immediately. Public opinion shifted dramatically against Germany. What German intelligence saw as a tactical victory became a strategic disaster.

The problems ran deeper than backlash. German spy networks suffered from poor tradecraft and basic security failures. One German officer, Franz von Papen (later Hitler's Vice Chancellor), left his briefcase containing spy contacts and payment records in a New York elevator. Another agent lost crucial documents when his mistress stole them during a lovers' quarrel and sold them to British intelligence.

These failures stemmed partly from Germany's intelligence structure. Unlike Britain's centralized approach in Room 40, German espionage in America operated through competing organizations that rarely coordinated efforts. Military attachés, civilian businessmen, and ethnic German-Americans all ran separate operations with conflicting priorities.

The Germans simply didn't understand America. They thought sabotage would scare Americans into staying neutral, but instead it convinced many that German militarism threatened American security directly.

The U.S. response changed American counterintelligence. Before 1916, the U.S. had minimal capability to track foreign agents. After Black Tom, the Bureau of Investigation (precursor to the FBI) expanded rapidly. New laws against sabotage and espionage were passed. Americans began seeing internal security as a national priority for the first time.

Lessons From World War I

These early intelligence operations—Hari's amateurish spying, Room 40's coding breakthroughs, Germany's American sabotage campaign—laid the groundwork for World War II.

Room 40's innovations would grow into Bletchley Park and the breaking of the Enigma code. The amateurish sabotage attempts in America would evolve into sophisticated deception operations like Operation Fortitude.

As war clouds gathered again in the 1930s, the world's powers remembered both the successes and failures of World War I intelligence. Those who paid attention and learned from the past had an advantage in the conflict ahead.

World War II: Intelligence Innovations and Strategic Impacts

The mansion sat in the quiet Buckinghamshire countryside. Its Victorian architecture hid the revolution happening inside. Behind its walls, an unlikely army fought a different kind of war—a war against codes the Germans believed were unbreakable.

They recruited people the military would never have considered. They put mathematicians beside linguists beside chess players and told them to think differently.

This tactic paid off against the fearsome Enigma machine, which we'll explore in more detail in the next chapter. With 159 million possible combinations, Enigma should have been undefeatable. The Germans thought exactly that, sending their most sensitive communications through a system they believed to be mathematically secure.

They were wrong. Through Polish mathematical insights, British engineering genius, and raw intellectual firepower, Bletchley Park began reading German messages. By 1943, Allied commanders often read German orders ahead of their German counterparts. This project was known as "Ultra."

It changed everything. Convoys carrying life-saving supplies changed course before waiting U-boats could strike. In North Africa, General Montgomery was aware of German General Rommel's plans, strengths, and fuel shortages. Before D-Day, Allied planners knew which German divisions were where and which beaches were most heavily defended.

But this advantage created a terrible dilemma. Using intelligence too obviously could alert the Germans that Enigma was broken. Military commanders faced impossible choices. Should they let a convoy attack to protect the secret? Should they let a bombing happen when they knew the target beforehand?

For every life saved by acting on Ultra intelligence, they risked millions by exposing their secret. Sometimes, they had to watch people die to protect the bigger secret.

The Double Game

Another British intelligence operation played an even more audacious game. The Twenty Committee, which was named for the Roman numeral XX, or "double cross," didn't just gather intelligence—they created it.

By 1942, British counterintelligence had captured every German agent sent to Britain. Instead of executing them, they offered them a choice: work for Britain or face the gallows. Most chose to live. These turned agents became the foundation of history's most successful deception operation.

Under British control, these double agents fed Germany a mix of truth and fiction. They reported accurate but minor information to build credibility, then delivered lies at just the right time. The Germans, believing they had reliable spies, based critical decisions on information manufactured in London offices.

The operation peaked before D-Day. Double agents convinced German high command that the main invasion would come at Calais, not Normandy. This scheme worked so well that even after troops landed in Normandy, Hitler held the most important divisions back, waiting for the "real" invasion at Calais.

They didn't just fool the enemy; they got inside their heads. They knew how they processed information, what they expected to hear, and how to make lies sound like truth.

The psychology was advanced. Each double agent had a carefully tailored personality and reporting style. Their handlers studied German intelligence analysts to understand what they found believable. The deceptions worked because they aligned with German expectations and biases.

From Lyon's Shadows to Interpol's Light

The opulent stone buildings along the Rhône River hide Lyon's darkest secrets. The French city of light and gastronomy has two faces—once the nightmare headquarters of Klaus Barbie, the Nazi "Butcher of Lyon," and now home to Interpol's gleaming command center where international police forces coordinate their hunt for modern criminals.

Walking these streets feels like crossing invisible timelines. The same winding passages where Resistance fighters once hid from Gestapo patrols now pulse with Interpol officers from 195 countries. The architectural contrast between Barbie's former torture chambers at Hotel Terminus and Interpol's sleek headquarters building tells a story of how far Lyon has traveled from its darkest hour.

Barbie in Lyon

In November 1942, a 29-year-old SS officer named Klaus Barbie arrived in Lyon to head the local Gestapo. His reputation had already been established in Amsterdam, where his brutal interrogation methods earned him his first promotion. Lyon would become the canvas for his most horrific work.

Barbie's specialty was breaking French Resistance networks that made Lyon their center of operations. He tackled this mission with ruthless sadism—mapping resistance cells, identifying key members, and then using torture to extract information that led to more arrests.

His most infamous act came in February 1943, when he arrested Jean Moulin, the highest-ranking member of the French Resistance. Moulin, who coordinated resistance activities across France, fell into Barbie's hands after being betrayed by a fellow resistance member. Barbie personally tortured Moulin for days, beating him, breaking his bones, and pulling his nails.

Moulin never talked and died from his injuries in transit to Germany, becoming France's most revered resistance martyr.

By summer 1943, Barbie's reputation had spread throughout occupied France. Captured resistance members knew what awaited them at 14 Avenue Berthelot, Barbie's headquarters. Some carried cyanide capsules to avoid the inevitable torture.

The building still stands today, repurposed as the Center for the History of the Resistance and Deportation—a museum that educates visitors about Lyon's dark history. Many of the exhibits include testimony from survivors of Barbie's torture chambers.

Barbie's brutality wasn't limited to resistance fighters. In April 1944, he orchestrated a raid on a children's home in Izieu, about 50 miles east of Lyon. The home sheltered Jewish children from across Europe. On Barbie's orders, Gestapo agents rounded up 44 children and their seven adult caretakers. All were deported to Auschwitz. None survived (Dawsey, 2019).

America's Deal With the Devil

As Allied forces neared Lyon in late August 1944, Barbie fled with other Nazi officials. What happened next is disturbing.

The Americans knew exactly who Barbie was and what he had done in Lyon. However, they made a calculated decision that his anti-communist clout was more valuable than justice for his victims.

In 1947, Barbie became an agent for the U.S. Army Counterintelligence Corps (CIC), working from Germany to recruit informants and monitor communist activities. When French authorities requested his extradition in 1950 to stand trial for war crimes, his American handlers helped him escape to Bolivia via a "ratline"—a Nazi escape route through Italy operated with the help of Catholic clergy.

For decades, Barbie lived openly in Bolivia, even as Holocaust survivors and resistance members recognized him. The Bolivian government, which benefited from his expertise in suppressing opposition, refused French extradition requests.

Bringing the Butcher to Account

In 1971, Nazi hunters Beate and Serge Klarsfeld tracked Barbie to Bolivia and launched a public campaign to expose his true identity. Their efforts led to increased international pressure, but it wasn't until 1983, after a change in the Bolivian government, that Barbie was finally extradited to France.

The trial took place in Lyon's Palace of Justice, just blocks from Barbie's former headquarters. He showed no remorse, explaining he was only following orders. After testimony lasting eight weeks, Barbie was convicted of crimes against humanity and sentenced to life imprisonment. He died of cancer in prison four years later.

The Showgirl Who Outsmarted the Third Reich

The Nazis never suspected that their downfall would come wearing a banana skirt. In Paris, 1940, German officers packed the front rows of the theater, mesmerized by Josephine Baker's legendary performances. Little did they know, they were watching one of Allied intelligence's most valuable assets in action. She wasn't just another entertainer—she was a masterclass in espionage hiding behind sequins and spotlights.

Baker's journey from St. Louis chorus girl to international spy reads like fiction. Fleeing Jim Crow America in the 1920s, she dominated Paris with her exotic dancing and magnetic charisma. When war erupted, French counterintelligence chief Jacques Abtey approached her with a proposition: would she spy for her adopted homeland?

Her response was unflinching: "The Parisians gave me their hearts, and I am ready to give them my life " (Klein, 2021).

What made Baker the perfect operative was her celebrity passport. She toured neutral Portugal and Spain carrying intelligence reports written in invisible ink on her sheet music. Border guards, starstruck by her fame, never thought to check the underwear where she pinned classified documents. One officer reportedly spent their "inspection" asking for an autograph instead.

Her château in the Dordogne became spy central—a place where she hosted elaborate parties for German officers, plying them with wine while extracting military secrets between champagne toasts. In Morocco, Baker performed for packed houses while secretly meeting Allied commanders, passing intelligence that shaped the North African campaign. The thunderous applause that followed her performances conveniently masked her covert activities.

For her extraordinary bravery, General de Gaulle personally decorated Baker with France's highest military honors. Unlike the tragic fate of many wartime spies, she lived to see the liberation she helped secure, later fighting another battle for civil rights with her "Rainbow Tribe" of adopted children from around the world.

Baker's genius lay in turning her greatest liability—being a highly visible Black female celebrity in Nazi-occupied Europe—into her greatest espionage asset. Sometimes, the best disguise is no disguise at all.

Reflections

Looking back at intelligence operations during World War I and World War II, it's striking how these covert campaigns changed history. These weren't minor sideshows—they were decisive turning points that saved thousands of lives and shortened conflicts.

The true stories are better than fiction. Mata Hari wasn't exactly who we thought she was. The quiet heroes at Room 40 and Bletchley Park broke "unbreakable" codes with basic tools and brilliant minds. Can you imagine deciphering the Zimmermann Telegram or cracking Enigma without computers? Pure genius.

What's fascinating about these accounts is how ordinary people accomplished extraordinary things with creativity and courage. When you understand these stories, you'll never look at wartime victory the same way again. The secret battles behind the visible war might be the most compelling stories of all.

Thank You for Reading

I'm grateful you chose to spend your time with THE SPY ARCHIVE. Every reader brings these stories to life in their own way, and your perspective matters.

If you enjoyed the book - or even if there's something you think could make it better - I'd love to hear from you. Your review helps other readers discover the book and supports the mission behind it.

Scan the QR code to leave your thoughts.

All proceeds from this book go to IN-Network.org, a nonprofit dedicated to inspiring and guiding the next generation of national security leaders.

Thank you for being part of this journey and helping us make a difference.

Chapter 7:

Unsung Heroes of WW2

History remembers the spectacle of war—the generals on horseback, the infantry charging, the thundering artillery. But intelligence work happens behind closed doors. Its greatest victories are the ones that history never records: the attack that didn't happen, the battle that was avoided, and the lives that were never put at risk.

This invisibility comes at a price. When history forgets its intelligence heroes, it often forgets those who society already pushed to the margins—the outsiders whose very "otherness" made them valuable to intelligence services even as it made them vulnerable to erasure afterwards.

The individuals we introduced in the previous chapter include James Lafayette, an enslaved man who risked everything to spy for the Continental Army during the American Revolution, and Harriet Tubman, who not only helped people escape through the Underground Railroad but also led Union intelligence operations behind Confederate lines. The very societies that oppressed them never suspected they would become their most effective intelligence assets.

Their stories reveal a painful paradox: Intelligence agencies often relied on those whom mainstream society rejected. The mathematician labeled "deviant" for his sexuality. The Native Americans punished for speaking their mother tongues. The women told their minds weren't capable of analytical work. The immigrants whose accents and cultural knowledge were needed for operations but whose loyalties were constantly questioned.

In wartime, desperation forced intelligence agencies to tap into these overlooked talents. In peacetime, those same agencies often discarded them. Their contributions were classified, and their sacrifices went unacknowledged.

This chapter focuses on those who served at the margins, how their differences became strategic advantages, and why their stories remained buried for decades. It discusses how prejudice shapes not just who we remember, but how we understand intelligence itself—and what we lose when we allow discrimination to dictate which heroes history honors.

These are the stories of those who saved nations that wouldn't save them in return—the codebreakers, spies, and analysts whose greatest achievement wasn't just outsmarting the enemy, but overcoming the barriers their own societies placed before them.

Their intelligence work changed history. Now, history must be intelligent enough to remember them.

Erased By Prejudice

On a cold morning in 1941, a lanky mathematician named Alan Turing hunched over papers in Hut 8 at Bletchley Park. German U-boats were winning the Battle of the Atlantic, sinking Allied ships faster than they could be built. Britain faced starvation. The Nazi naval commanders coordinated these devastating attacks using messages encrypted by what they believed was an unbreakable system: the Enigma machine.

The Enigma machine looked like an oversized typewriter, but it was actually an engineering marvel. Each keypress passed through a series of rotating wheels and electrical connections, appearing as a different letter. The wheels' positions changed after each letter, creating a mind-boggling code.

What the Germans hadn't counted on was Turing's extraordinary talents. Where others saw an insurmountable mathematical barrier, Turing saw patterns and possibilities.

Turing designed electromechanical machines called "bombes" that could rapidly test possible Enigma settings. But the real breakthrough came from looking for patterns and predictable content—what cryptographers call "cribs." Cribs could be weather reports sent at the same time each day, messages that always ended with "Heil Hitler," or even ships using standard formats to report their positions (Cox, 2018).

The results were extraordinary. By late 1942, Bletchley Park was regularly reading German naval communications, giving Allied convoys useful information about U-boat positions. It's estimated that Turing's work shortened the war by two to four years.

After such monumental achievements, you might expect Turing would have been celebrated as a national hero. Yet for decades after the war, his contributions remained virtually unknown to the public. The bitter truth is that intelligence often relies on the marginalized, only to discard them afterward.

In 1952, just seven years after helping win the war, Turing was arrested for homosexual acts, then illegal in Britain. Given a choice between prison and chemical castration, he chose hormone treatments that caused physical and psychological suffering. His security clearance was revoked. Two years later, at just 41, he died from cyanide poisoning—widely considered suicide.

Through the 1950s and '60s, while veterans of conventional warfare received medals and parades, Turing's name remained virtually unknown. The British government's Official Secrets Act kept Bletchley Park's achievements classified until the 1970s. Even then, Turing's contributions weren't fully recognized.

Britain finally addressed this historical wrong in 2009 with a public apology to Turing, followed by a royal pardon in 2013. In 2017, a law known as "Turing's Law" pardoned thousands of gay men convicted under historical legislation. His face now appears on the £50 note.

The Navajo code talkers were another disadvantaged group who used their difference to save countless lives.

The Forbidden Tongues that Won the War

A Japanese intelligence officer scrutinized a recorded transmission, having listened to the tape for the 20th time. The unique, flowing language with its distinctive tone was hard to figure out. His team of elite cryptographers had cracked every American code—except this one. To Imperial Japan's finest minds, it sounded like gibberish, but it was actually America's most devastating secret weapon.

It wasn't a machine. It wasn't mathematics. It was a language that the American government had once tried to exterminate.

The year was 1942, and Philip Johnston—a World War I veteran raised on a Navajo reservation—approached Marine commanders with an audacious proposal. America needed an unbreakable code for Pacific operations. Johnston knew exactly where to find it: among the very people whose languages federal boarding schools had beaten children for speaking.

The Navajo language was perfect for military encryption. It contained sounds no Japanese linguist could replicate, used sounds that made transcription virtually impossible, and existed almost entirely within the American Southwest. With no published dictionaries or learning materials, enemy intelligence had zero reference points.

The initial 29 Navajo recruits didn't just translate messages—they engineered an elaborate military vocabulary from scratch. Planes became "birds." Fighters were "hummingbirds," and bombers became "buzzards." Tanks were "turtles," submarines were "iron fish," and America itself was "our mother." They couldn't just say "tank" in Navajo because there's no word for that. So, they called it "*chay-da-gahi*": "tortoise-crawler."

At Iwo Jima, where Marines suffered 26,000 casualties, six Navajo code talkers worked around the clock, transmitting over 800 flawless messages in the first 48 bloody hours of battle. Were it not for the Navajos, the Marines would never have taken Iwo Jima.

The irony of history is that these men volunteered to protect a nation that systematically tried to eliminate their languages. Many Code Talkers had childhood scars from government teachers who washed their mouths with soap for speaking the very words now saving American lives.

Their extraordinary service remained classified until 1968. Most returned to reservations where they couldn't vote until 1948. The original Navajo code talkers weren't awarded Congressional Gold Medals until 2001, after most had already died.

These Native Americans turned persecution into patriotism. Turning the language America tried to destroy into the code America couldn't survive without.

Women of Intelligence

In June 1943, a German security alert went out across occupied France: "The woman who limps is one of the most dangerous Allied agents in France. We must find and destroy her" (Neven, 2018).

That woman was Virginia Hall, an American spy with a wooden leg she nicknamed "Cuthbert." The Gestapo called her "the limping lady" and considered her so dangerous that they launched a nationwide manhunt to capture her. She escaped by hiking over the snow-covered Pyrenees Mountains on her prosthetic leg, continuing her intelligence work that helped pave the way for D-Day.

Hall's story is just one of thousands that were buried in classified files for decades. While male war heroes received medals and movie deals, the women of World War II who worked in intelligence often operated in obscurity that didn't lift even after peace came.

Female intelligence officers often kept their wartime roles a secret for the rest of their lives. Many returned to civilian life without ever telling family members what they had done during the war. The intelligence community, still dominated by men in the postwar years, rarely acknowledged their contributions publicly.

The Princess Who Became a Spy

Few stories are as fascinating—or as tragic—as that of Noor Inayat Khan. The daughter of an Indian Sufi mystic and an American mother, Noor was a Muslim and the first woman Special Operations Executive (SOE) agent to parachute into France.

She was raised in a pacifist household where non-violence was a sacred belief. Yet when Nazi forces occupied France, this gentle woman who once wrote children's stories became one of the most important—and courageous—agents in Britain's fight against fascism.

Noor defied every stereotype of what a spy should be. She was shy, unworldly, and deeply spiritual. Her SOE trainers worried she was too honest and kind-hearted to handle secret operations. They couldn't have been more wrong about what she was capable of.

The SOE was Britain's sabotage and subversion agency, tasked with "setting Europe ablaze" through resistance activities. By 1943, SOE faced a critical shortage of wireless operators willing to parachute into occupied France—a virtual death sentence with an average survival time of just six weeks. The Germans had become extraordinarily efficient at tracking radio transmissions and capturing operators.

Noor, fluent in French from her Parisian upbringing and trained in radio operations, volunteered despite knowing the odds. On a moonlit night in June 1943, she parachuted into occupied France under the codename "Madeleine" with a life expectancy measured in weeks, not months.

Noor landed in the Loire Valley and made her way to Paris, where she joined the "Prosper" network. Just days after her arrival, disaster struck. The Gestapo arrested nearly every member of the network in a devastating sweep. Noor's commanding officer offered her the chance to return to England.

She refused.

For the next four months—far longer than the six-week average—Noor worked as the last remaining radio operator in Paris, constantly moving her equipment between safe houses, transmitting vital information back to London. She evaded capture repeatedly, sometimes escaping just minutes before Gestapo teams arrived.

Her luck ran out in October 1943 when a Frenchwoman betrayed her for 100,000 francs. The Gestapo arrested her at her apartment, finding her codebooks, which she had been ordered to destroy but had kept in hopes of memorizing them and continuing her work.

What followed showed the steel beneath Noor's soft exterior. Imprisoned at Avenue Foch, Gestapo headquarters, she made two escape attempts. During one, she almost succeeded, climbing onto the roof before being recaptured. Her interrogators, frustrated by her refusal to reveal information about SOE operations, classified her as "extremely dangerous."

In November 1943, the Nazis sent Noor to Pforzheim prison in Germany, where she was kept in chains and solitary confinement for ten months. Even her guards weren't allowed to speak to her. Throughout her imprisonment, she revealed nothing, not even her real name.

On September 13th, 1944, as Allied forces pushed into Germany, Noor and three other female SOE agents were taken to Dachau concentration camp. There, kneeling in the rain, they were executed with shots to the head. Witnesses reported that Noor was severely beaten before her execution but died with the word "Liberté" on her lips.

She was 30 years old.

The Women Who Won the War

The intelligence war of 1939–1945 created unprecedented opportunities for women. At Bletchley Park, they made up a large percentage of the workforce. At Arlington Hall, where Japanese codes were broken, the figure was high. The Office of Strategic Services (OSS)—America's first spy agency—recruited female professors, journalists, actresses, and socialites for dangerous field missions.

Their contributions came in every aspect of intelligence work:

- **In field operations:** Aline Griffith, Countess of Romanones, who kept her cover as a fashionable socialite in Madrid while running a spy network that exposed Nazi smuggling operations through neutral Spain (Jureidini, 2024).

- **In analysis:** Elizabeth Friedman, who not only broke enemy codes but also testified in court against Nazi spies, helping convict 33 German agents.

- **In resistance:** Pearl Witherington, who took command of a 3,500-strong French Resistance network after the Germans captured her male colleague, and was so effective that the Germans put a one-million-franc price on her head.

These weren't supporting players but central figures whose work directly saved thousands of lives and shortened the war. Yet the barriers they faced were extraordinary.

Fighting Two Enemies

Women in intelligence fought two battles simultaneously: one against the Axis powers and another against institutional sexism.

The War Department initially opposed employing women in intelligence roles, believing them unfit for such work. Many women faced pay discrimination, were classified in lower professional categories despite doing identical work to men, and struggled to receive proper equipment and training.

In Britain, female SOE agents parachuting into occupied territory received smaller weapons than men—not because they requested them, but because male commanders assumed women couldn't handle full-sized firearms. Many found ways to trade up after landing, knowing their lives depended on the right gear.

At Arlington Hall, brilliant female cryptanalysts who made critical breakthroughs with Japanese codes were classified as "sub-professional" employees, while men with far less responsibility were labeled "professional." The gap in status and pay persisted despite equal or superior performance.

Yet these women fought on, motivated by patriotism and the rare chance to use their full intellectual abilities. The desperate need for results sometimes overrode prejudice.

Hiding in Plain Sight

Many female agents succeeded precisely because enemies shared the same prejudices as their superiors. The Nazis and Japanese often overlooked women as potential threats.

Female agents realized that being underestimated was their greatest advantage. German patrols would walk past women carrying military plans while stopping men for routine searches. This blind spot in enemy security became a critical vulnerability that Allied intelligence exploited repeatedly.

Female spies turned stereotypes into weapons. They smuggled explosives in grocery bags, hid receivers in sanitary products (knowing male guards wouldn't check), and passed messages while getting their hair done—activities too "feminine" for enemy counterintelligence to closely monitor.

This talent for hiding in plain sight saved countless lives. When Marie-Madeleine Fourcade's French resistance network was compromised, she escaped prison by stripping naked, becoming so thin without clothes that she could squeeze between her cell bars. She went on to provide intelligence that mapped every German defensive position on the Normandy coast before D-Day.

The Silent Legacy

After the war ended, most women in intelligence faced a difficult transition. With men returning from combat, women were pushed out of their positions. They were expected to return to domestic life without ever speaking about their wartime accomplishments.

Some persisted anyway. Elizabeth Friedman continued her codebreaking work at the newly formed National Security Agency (NSA). Marion Frieswyk became the CIA's first female intelligence analyst. Virginia Hall, despite her disability and facing continued discrimination, worked for the CIA until 1966.

For most, however, their contributions remained unacknowledged for decades. Many died without public recognition, their families never knowing what they had actually done during the war.

The secrecy that protected their work became a barrier to recognition. But as classified files opened in recent decades, these women's stories have finally come to light. Their stories reveal that much of what we thought we knew about World War II intelligence was incomplete.

These weren't just women who served during wartime. They were pioneers who changed the very nature of intelligence and warfare, proving that gender meant nothing compared to courage, creativity, and commitment.

Reflections

In June 2013, a small ceremony took place at Britain's Government Communications Headquarters (GCHQ). Officials unveiled a simple plaque bearing Alan Turing's name and the dates of his birth and death. The location was fitting—GCHQ is the direct descendant of Bletchley Park's codebreaking operation, where Turing helped defeat Hitler's Germany.

The plaque came 60 years too late for Turing himself. It arrived decades after his security clearance was revoked, after his forced chemical castration, after his suicide at age 41. It couldn't undo the suffering inflicted on a man who had saved his country, only to be persecuted by it.

This pattern—of intelligence agencies exploiting marginalized people's unique talents during crises, then discarding or persecuting them afterwards—appears repeatedly throughout intelligence history. The narrative isn't simply about ingratitude; it reveals how prejudice undermines national security itself.

Consider what was lost when Virginia Hall was repeatedly passed over for CIA leadership positions despite her extraordinary wartime accomplishments. How many operations might have benefited from her expertise? What insights went untapped because male superiors couldn't imagine a woman with a disability leading intelligence operations?

Or consider the decades when Native American languages were suppressed in government boarding schools, even after Navajo code talkers proved their strategic value. How many other potential strategic assets did American intelligence lose through cultural blindness?

These weren't just moral failures but strategic ones. Intelligence agencies that dismiss talent because it comes in "unexpected packages" don't just harm individuals—they harm national security by narrowing their own capabilities.

Yet, history also offers hope. When desperation forced intelligence agencies to overlook prejudice and recruit based solely on talent, extraordinary innovations emerged. Bletchley Park's gender-integrated teams pioneered computing techniques that transformed the modern world. The OSS's recruitment of immigrants, women, and minorities created operational capabilities no homogeneous agency could match.

These successes hint at what intelligence services might achieve if they embraced diversity not just during crises but as a standard practice. Recent decades have shown promising shifts: Intelligence agencies now actively recruit from previously marginalized communities, recognizing that diverse perspectives strengthen analysis and operations.

The CIA's Center for Mission Diversity and Inclusion, established in 2010, represents an institutional acknowledgment that homogeneity creates blind spots. GCHQ has publicly apologized for its historical treatment of LGBTQ+ employees and now actively recruits from those communities. Female directors have led major intelligence agencies in multiple countries.

But the most radical change may be in how we tell the history of intelligence. As classified files open and historians look beyond traditional sources, a more complete picture emerges—one that finally acknowledges contributions previously erased by prejudice.

Josephine Baker and Noor Khan now appear in French intelligence histories alongside male heroes of the Resistance. Elizabeth Friedman's cryptanalytic achievements garner scholarly attention equal to her husband's. The Navajo code talkers are celebrated in museums, books, and schools.

This rewriting of intelligence history isn't just about recognition, though recognition matters deeply to those previously erased. It's about understanding intelligence itself more accurately. The lone male genius operating independently was never the full story. Intelligence success has always relied on diverse teams bringing different perspectives, skills, and backgrounds to impossibly difficult problems.

As intelligence agencies face 21st-century challenges—from cyber threats and global terrorism to climate security and pandemic early warning—they need this diversity more than ever. The next Turing may speak with an accent, have a different gender identity, practice an unfamiliar religion, or move through the world in a different way. Intelligence services that recognize this talent, regardless of its form, will have the advantage.

Chapter 8:

Shadows of the Cold War

When World War II ended in 1945, a new conflict began before the celebration parades had even finished. This war would last forty years, span the globe, and redefine espionage forever. The Cold War pitted two intelligence titans against each other in a high-stakes game where nuclear annihilation was always one miscalculation away.

On one side stood the newly formed CIA, born in 1947 from the ashes of the wartime OSS. On the other side was the feared KGB, created in 1954 as the latest version of Soviet intelligence. Their battle for information, influence, and advantage altered the course of history.

CIA and KGB

When the guns of World War II fell silent in 1945, another conflict played out behind closed doors—not with tanks and bombers, but with secrets and lies. The Cold War created the modern intelligence agency, transforming espionage from a wartime necessity into a permanent institution of state tradecraft and power.

America Creates a Peacetime Spy Agency

In 1947, President Truman signed the National Security Act, giving birth to the Central Intelligence Agency. This decision marked a radical departure for America, which had historically dismantled its intelligence capabilities after each war. The CIA emerged from the ashes of the OSS, but with one major difference—it would be the first permanent American peacetime intelligence service.

The early CIA faced skepticism from all sides. Congress feared creating an American Gestapo. The military and the FBI viewed the newcomer as a bureaucratic rival.

Allen Dulles, who became the agency's longest-serving director, shaped the CIA's identity during these formative years. A Princeton-educated lawyer and OSS veteran, Dulles envisioned an intelligence service that could fight communism through both information gathering and direct action. Under his leadership, the CIA expanded beyond traditional intelligence into covert operations—overthrowing governments in Iran and Guatemala and establishing a global network of agents and assets.

The CIA developed a distinctive character that reflected American culture and values. It recruited heavily from Ivy League universities, created technological solutions to intelligence problems, and operated with bureaucratic caution and operational boldness. CIA officers saw themselves as the front line against Soviet expansion, defending democracy through methods democracy itself sometimes couldn't accept.

The Soviet Response

The Soviets were working from an entirely different playbook. They already had decades of experience running spy networks dating back to the Russian Revolution. After several reorganizations, Soviet intelligence became the KGB in 1954—a massive organization with far more power than its American counterpart.

The KGB didn't just spy on foreigners; it watched Soviet citizens, guarded borders, protected government officials, and hunted political opponents. Where the CIA faced legal limits and congressional oversight, the KGB answered only to top Communist Party leadership. One KGB chairman, Yuri Andropov, eventually became the Soviet leader.

The KGB organized itself like a military operation. Different directorates handled specific tasks—foreign intelligence, counterintelligence, border guards, and government security. This structure created specialists who mastered their particular field.

Their training was legendary. Future spies spent years learning languages, memorizing foreign city layouts, studying psychology, and practicing the art of recruiting agents. The Americans had technology. The Soviets had people who understood people.

Two Models of Intelligence

As these giants squared off, they developed completely different styles of intelligence work.

The CIA fell in love with technology. They built the U-2 spy plane that could photograph Soviet bases from the edge of space. They created satellites that could read newspaper headlines from orbit. They set up listening posts to capture Soviet radio traffic. This technology-first approach fits American strengths: industrial might and scientific innovation.

The KGB bet on human intelligence. They excelled at finding people with access to secrets and convincing them to share those secrets— sometimes through ideology, sometimes through money, sometimes through blackmail. They played a patient game, sometimes nurturing potential agents for years before making a pitch.

You can see these differences in how they operated. The CIA might spend millions on a satellite to photograph a Soviet missile base. The KGB would spend years getting someone who worked at that base to smuggle out the technical manuals. Both methods worked, but in completely different ways.

Spies Face Off Around the World

Berlin became a spy hub during the Cold War. The divided city created an unusual environment where East and West stared at each other across a wall. Both sides packed Berlin with intelligence officers, creating a breeding ground for espionage.

The CIA set up shop in West Berlin, running operations into East Germany and beyond. The KGB and its East German partner, the Stasi, made West Berlin a prime target for recruitment and surveillance. In this pressure cooker, mistakes could be fatal. Double agents, defections, and kidnappings became regular occurrences.

But the spy war spread far beyond Berlin. In Vienna, CIA and KGB officers played cat-and-mouse through the city's famous sewer system. In Afghanistan, the CIA supplied weapons to fighters battling Soviet forces. In Cuba, the KGB helped Castro spot and neutralize CIA-backed rebels. From Africa to Latin America to Asia, both agencies turned local conflicts into proxy battles.

There were plenty of failures on both sides. The CIA's Bay of Pigs invasion of Cuba in 1961 became a public embarrassment. This disastrous operation involved approximately 1,400 CIA-trained Cuban exiles attempting to overthrow Fidel Castro's government. Poorly planned and executed, the invasion force was quickly defeated by Castro's military, resulting in nearly all invaders being killed or captured. The fiasco damaged America's international reputation and strengthened Castro's position while pushing Cuba further into alliance with the Soviet Union. The KGB's heavy-handed control of Eastern Europe led to resentment that would later help bring down the Soviet system.

But both agencies scored major wins, too. The CIA's recruitment of Soviet Colonel Oleg Penkovsky gave America critical intelligence during the Cuban Missile Crisis. The KGB's penetration of NATO security provided Soviet leaders with insights into Western military planning. Each success forced the other side to up its game.

What They Left Behind

When the Soviet Union fell apart in 1991, both spy services faced identity crises. The KGB was dismantled, and its foreign intelligence work was handed over to a new agency called the SVR. The CIA, suddenly without its main adversary, searched for new missions fighting terrorism, narcotics, and rogue states.

But the legacy of their 40-year duel lives on in how intelligence works today. The techniques they developed—from satellite surveillance to cyber operations to influence campaigns—still form the backbone of modern espionage. The organizational structures they created still shape intelligence agencies worldwide.

Think about how intelligence appears in movies, books, and news. Most of those images—the spy with a hidden camera, the satellite zooming in on a secret facility, the defector smuggled across a border—come directly from Cold War operations that the CIA and KGB pioneered.

The CIA-KGB rivalry changed the world of espionage, laying the groundwork for what modern intelligence looks like today. They built something never seen before in history: massive, permanent, global intelligence organizations that operate in both wartime and peace. That's what makes their story so fascinating—and so important for making sense of today's headlines about Russian hackers, CIA operations, and global conflict.

MI6 and Space Espionage

Beyond changing the CIA and KGB, the Cold War forced British intelligence to rebuild itself almost from the ground up. While American and Soviet spy agencies expanded, MI6 faced a different challenge: recovering from the worst spy scandal in its history while adjusting to Britain's diminished global role.

This story shows how an intelligence service can bounce back from disaster and find new purpose in a changing world.

The Cambridge Spy Ring

In 1951, two British diplomats suddenly disappeared. Donald Maclean and Guy Burgess had fled to Moscow, exposing what would become known as the Cambridge Five—a Soviet spy ring that had infiltrated the heart of British intelligence.

Three more men were eventually identified as Soviet agents: Kim Philby (who had been in line to head MI6), Anthony Blunt (who worked in counterintelligence), and John Cairncross (who had access to atomic secrets). All were recruited at Cambridge University in the 1930s and had risen to positions of extraordinary access.

The damage was catastrophic. Philby alone had compromised hundreds of operations and gotten dozens of Western agents killed. Maclean had passed nuclear secrets to the Soviets, helping accelerate their atomic bomb program. For years, Moscow had known virtually everything about British (and many American) intelligence operations.

M16 wasn't just defeated—they were fooled. The people they trusted most were working against them the whole time. The scandal shattered morale and destroyed MI6's reputation with its allies. The CIA became reluctant to share information, fearing more leaks. Inside MI6, officers suspected colleagues of being undiscovered spies. The organization that had once run one of the world's most effective intelligence networks now struggled with basic operational security.

Rebuilding Trust and Finding New Purpose

MI6 faced hard questions about its future. Britain was no longer a superpower. Its empire was crumbling. Scandals had exposed fundamental weaknesses in recruitment and security. Some politicians wondered whether Britain even needed an intelligence service of its size.

The answer came through reinvention. MI6 painstakingly rebuilt itself with new security protocols, more thorough background checks, and a focus on areas where Britain could still contribute meaningfully to Cold War intelligence efforts.

The special relationship with America became central to this strategy. MI6 had something the CIA needed: experience. Despite the Cambridge disaster, British officers still had unmatched expertise in running agents, analyzing intelligence, and conducting covert operations. Many had decades of experience from World War II and colonial conflicts.

Dick White, who took over MI6 in 1956, focused on restoring this relationship. They couldn't match America in money or technology, but they could still teach them about human intelligence.

The strategy worked. By the early 1960s, MI6 and the CIA had developed what both sides called the "cousins" relationship—an unusually close partnership that involved sharing intelligence, coordinating operations, and even jointly running agents.

This partnership helped both agencies. The CIA gained access to British expertise and global networks. MI6 gained access to American resources and technology. Together, they created an intelligence alliance that the Soviet bloc struggled to match.

Looking to the Stars

As the Space Age began, intelligence gathered from satellites and space-based systems became increasingly important. The Soviets' launch of Sputnik in 1957 created new urgency—if they could put a satellite in orbit, they could potentially put nuclear weapons in orbit, too.

MI6, lacking the resources to build its own satellite program, made a strategic decision to concentrate on analyzing data rather than collecting it. They excelled at interpreting information gathered by American satellites, adding context and identifying patterns that might otherwise be missed.

British scientists and intelligence analysts worked closely with their American counterparts to understand Soviet space capabilities. This collaboration expanded to monitoring Soviet missile tests, tracking space launches, and assessing new weapon systems.

The Americans had the cameras in the sky. They helped them figure out what they were looking at.

This focus on space intelligence allowed MI6 to make significant contributions despite Britain's limited resources. It also created knowledge that proved valuable in understanding other technological threats, from nuclear weapons to advanced aircraft.

The Evolution of Espionage Technology

"They found one in a bowl of caviar, you know," said the aging CIA technician, his eyes twinkling nostalgically. "Planted it right there among the eggs at a diplomatic reception in Prague. Recorded seven hours of Soviet trade negotiations before someone ate it by mistake." Though this exchange is fictional, the sentiment behind it is not. The Cold War's invisible battlefield was populated by more than trench-coated spies exchanging briefcases on foggy bridges. It was filled with ingenious devices that could hear through walls, photograph documents in darkness, and transmit secrets across continents. These technological marvels—many of which were declassified only in recent decades—reveal a parallel history of scientific innovation driven not by the market or academia, but by the desperate need to know what the other side was planning.

The Martini Olive and Other Miracles

The miniaturization race kicked off. Teams of engineers—many recruited directly from top universities before they could be snapped up by private industry—worked in secret labs to create listening devices smaller than anyone thought possible.

In 1956, the KGB created what technicians nicknamed "the Thing"— an almost magical listening device with no batteries or active electronic components. It was concealed inside a wooden carving of the Great Seal of the United States and presented to the American ambassador in Moscow. It remained undetectable to standard bug sweeps because it only activated when hit with a specific radio frequency from a Soviet van parked outside the embassy.

When the CIA discovered how it worked, they couldn't even be properly angry. They envied its brilliance.

The Soviets weren't the only ones with tricks up their sleeves. Western agencies developed cameras so small they fit inside coat buttons, recording devices thinner than a playing card, and microphones that could pick up sounds from hundreds of feet away when aimed at a window (because glass vibrates minutely when people speak, creating recoverable sound patterns).

MI6 technicians created a particularly clever device disguised as a pack of cigarettes. When the agent pressed a hidden switch, it would emit a puff of real cigarette smoke—the perfect cover while photographing documents with the pinhole camera concealed inside.

These gadgets weren't just technologically impressive—they were often beautiful expressions of craftsmanship. KGB technicians created aerated coins so perfect that they passed through circulation without detection. CIA disguise experts created false scars, moles, and even dental work that could transform an agent's appearance for a single critical meeting.

The Tunnel That Changed History

Sometimes, the most audacious technical operations were also the largest. None exemplifies this better than the Berlin Tunnel of 1954.

Imagine trying to dig a quarter-mile tunnel under enemy territory, in complete secrecy, while removing enough dirt to fill a swimming pool—and doing it all without making noise or being spotted from above. Now imagine doing all that while knowing that getting caught could trigger an international crisis.

The tunnel, jointly operated by the CIA and MI6, tapped into Soviet communication lines, yielding over 50,000 hours of recorded conversations. Engineers developed special recorders that minimized tape changes, air conditioning systems that wouldn't leave telltale condensation, and sound-dampening techniques to prevent detection.

What makes this story particularly fascinating is its double nature: George Blake, a Soviet mole in MI6, informed Moscow about the tunnel before construction even began. The Soviets, making a brilliant strategic decision, allowed the operation to continue while carefully controlling what information passed through those phone lines.

It became a strange psychological game. The Soviets knew they were listening, they didn't know they knew, and they couldn't react too obviously without revealing their source. So, this massive technological achievement became a stage for an even more intense human drama.

When the Soviets finally "discovered" the tunnel during staged road work in 1956, both sides had already achieved what they wanted: the West had gathered valuable intelligence, and the Soviets had protected their valuable agent while learning about Western technical capabilities.

Deadly Umbrella Tips and Poison Pens

Not all Cold War gadgets were designed to collect information. Some had darker purposes.

In 1978, Bulgarian activist Georgi Markov was waiting for a bus on London's Waterloo Bridge when he felt a sharp pain in his leg. A man nearby mumbled "sorry" and hurried away, carrying an umbrella despite the clear weather. Four days later, Markov was dead—assassinated with a ricin-filled pellet fired from that specially modified umbrella.

The KGB's "special techniques" laboratory developed an array of assassination tools: poisons that mimicked heart attacks, cameras that could fire lethal darts, and even a cigarette case that dispersed an incapacitating gas. Most remained thankfully unused, but their very existence pushed Western agencies to develop their own defensive technology.

CIA technicians created secret compartments in shoes, belt buckles large enough to hide escape maps but small enough to pass inspection, and false passports so convincing they could withstand expert scrutiny. One particularly ingenious device was the "rectal tool kit"—a small cylinder containing lock picks and escape tools that an agent could conceal internally if captured.

They had to assume the worst-case scenario for every operation. What if they're caught? What if they're searched? What if they need to escape in the middle of the night? Technology had to have an answer for each of these questions.

The Arms Race of the Airwaves

While miniaturized gadgets grabbed the public's attention, the true technological battle of the Cold War unfolded across the electromagnetic spectrum.

The National Security Agency and its British counterpart, Government Communications Headquarters (GCHQ), built thousands of radio receivers that could intercept communications from thousands of miles away. Soviet military units, diplomats, and spies all relied on radio transmissions, creating an invisible ocean of information for spies.

At RAF Menwith Hill in northern England, massive golf ball-like structures housed sensitive receivers that could pluck Soviet military communications from the air. In remote locations from Alaska to Turkey, listening posts operated around the clock. These posts were staffed by linguistic specialists who could distinguish between Russian military dialects or recognize individual Morse code operators by their distinctive "fists" (sending styles).

The Soviets knew they were listening, of course. But what could they do? The Americans couldn't stop communicating. Instead, they created layers of codes and ciphers, which the Americans and British then worked to break, which prompted the Soviets to create new ones. It was like an endless dance where neither partner could leave the floor.

The technical challenge drove innovations in computer technology. Early computers like the massive vacuum-tube ENIAC were first used to crack Soviet codes. As computing power increased, so did the sophistication of encryption and decryption methods, pushing forward the boundaries of what machines could do.

The Spy Satellites That Changed Warfare

Perhaps no technology transformed intelligence more fundamentally than spy satellites. The first successful American photo reconnaissance satellite, CORONA, began operating in 1960. By the time the program ended in 1972, CORONA satellites had taken over 800,000 images covering 750 million square miles of territory.

These satellites used film cameras—digital technology wasn't yet advanced enough. After taking pictures, the satellites ejected capsules containing film, which literally fell from space. Air Force planes equipped with special hooks would catch these capsules in mid-air as they descended by parachute—a technique so difficult it was compared to "catching a bullet with a bullet." Before satellites, they were blind. They would hear rumors about new missile sites or bomber bases, but couldn't confirm them without risking pilots' lives in overflights. Satellites changed everything. Suddenly, they could see what was really happening.

This visibility had far-reaching effects beyond intelligence gathering. It actually helped prevent conflict by reducing doubt. When Soviet leaders knew their actions were visible from above, the temptation to make secret military moves decreased. Similarly, American policymakers could verify that threatening Soviet activities were not underway, preventing panic from false alarms.

The Human Touch in a Technical World

For all the emphasis on gadgetry, the most successful operations still required the human element. The best technical collection had to be paired with skilled analysis and operational judgment.

The CIA's Technical Services Division stressed this point in its training programs. Technology amplifies human judgment but cannot replace it. The cleverest listening device in the world is useless if placed in the wrong room. The most advanced camera is worthless if aimed at meaningless documents.

This fusion of human and technical intelligence created some of the Cold War's greatest intelligence successes. Colonel Oleg Penkovsky, a high-ranking Soviet military intelligence officer, worked with Western intelligence from 1960 to 1962. Armed with a miniature Minox camera, he photographed thousands of pages of classified Soviet documents on missiles, nuclear planning, and military organization. This technical capability, combined with Penkovsky's human judgment about which documents mattered, made the operation one of the most valuable intelligence coups of the Cold War.

During the Cuban Missile Crisis, this human-technical partnership proved decisive. Satellite imagery spotted the missile installations, signals intelligence detected revealing communications, and human sources confirmed Soviet intentions. Together, these intelligence streams gave President Kennedy valuable information during the most dangerous confrontation of the Cold War.

The Legacy

When the Berlin Wall fell in 1989, the Cold War's abrupt end left intelligence agencies with powerful technological capabilities and suddenly unclear missions. The technologies developed for tracking Soviet submarines were repurposed to monitor terrorist communications. Satellite systems designed to count missile silos were redirected to watch for nuclear proliferation in new countries.

Many innovations eventually made their way into civilian life. Advances in miniaturization contributed to the development of personal electronics. Satellite technology evolved into GPS navigation. Even the internet itself owes something to ARPANET, a communications system designed to survive a Soviet nuclear attack.

But the most enduring legacy isn't any specific device—it's the mindset that technology could provide answers to security questions. This faith in technical solutions continues to drive intelligence agencies today, sometimes at the expense of human considerations.

The story of Cold War spy tech isn't just about clever gadgets or engineering breakthroughs—it's about how necessity drives innovation, how technology shapes human decisions, and how the quest for information can both increase security and create new vulnerabilities.

The next time you unlock your phone with facial recognition or ask a smart speaker for information, remember: you're using the great-grandchildren of technologies born in the secret workshops of Cold War intelligence agencies, where creativity and paranoia combined to change the world in ways we're still discovering.

Reflections

The Cold War wasn't just a standoff—it was the Super Bowl of spying.

While nuclear missiles stood ready, the real action happened where CIA and KGB agents played a deadly game of hide-and-seek. Although they were officially agencies, they functioned more like competing empires, complete with their own rules, heroes, and tragedies.

It's fascinating how they constantly one-upped each other. When Americans built spy planes, the Soviets shot them down. The KGB planted bugs, so the CIA recruited moles. Every day brought new tricks: poison-tipped umbrellas, cameras hidden in coat buttons, secret writing that appeared only under special lights.

It's intriguing how they balanced old-school spycraft with sci-fi tech. One agent might be turning sources in smoky bars while another monitors satellite signals from space. Their ingenuity was off the charts! MI6 took us to the final frontier. British intelligence and actual space espionage! Talk about raising the stakes.

Chapter 9:

Masters of Deception—The Cold War's Most Influential Spies

The most dangerous battlefields weren't in Korea or Vietnam. They were in everyday places: park benches where microfilm changed hands, embassy basements where secrets were whispered, and suburban homes where America's most protected codes were sold over family dinners.

History remembers the speeches. The summits. The standoffs. But victory and defeat often hinged on acts so small they seemed insignificant. A camera click in a Moscow bathroom. A document slipped into a shopping bag. A midnight radio transmission from behind enemy lines.

These weren't the spies of Hollywood's imagination. There were no tuxedos, no martinis, and no glamorous casino showdowns. The real masters of espionage were accountants with grudges, clerks with access, and analysts with financial troubles. They were ordinary people making extraordinary choices with world-changing consequences.

The spymaster's true weapon wasn't the poison umbrella or the hidden gun, but something far more powerful: understanding human weakness. Every major intelligence breach began with someone who could be compromised by money, ideology, coercion, or ego.

The Double Agent's Dangerous Game and Aldrich Ames Case Timeline

In the twilight hours of April 16th, 1985, Aldrich Ames walked into the Soviet Embassy in Washington with an envelope containing the names of three CIA assets inside the USSR. By nightfall, these men were marked for death, and Ames had crossed a line from which there was no return. The KGB paid him $50,000 on the spot—the first installment of what would become a $4.6 million betrayal (Case Study, n.d.).

History lovers have long been fascinated by double agents—those rare individuals who navigate two opposed worlds simultaneously, keeping false loyalties so convincingly that both sides believe they're seeing the authentic person.

Their stories provide a unique lens into the dark corners where personal psychology collides with world-changing events.

Ames's betrayal devastated American intelligence operations. A CIA veteran with access to the agency's most sensitive Soviet operations, he exposed virtually every significant human asset the U.S. had developed inside the Soviet Union. The human toll was immediate and brutal. At least ten CIA sources were executed, among them some of America's most valuable intelligence assets. General Dmitri Polyakov, who had provided important information on Soviet nuclear capabilities for decades, was arrested in 1986 and executed two years later.

What drove Ames to such treachery? Unlike ideological traitors of earlier eras, his motivation was disarmingly simple: money. Drowning in debt after an expensive divorce and struggling to maintain his new Colombian wife's lavish lifestyle, Ames made a coldly transactional decision.

He didn't betray his country for communism. He betrayed it for a Jaguar and a house in Arlington. This venal motivation is a striking contrast to that of Ames's most infamous predecessor, Kim Philby, the British intelligence officer who spied for the Soviets from the 1930s until his dramatic 1963 defection to Moscow.

Kim Philby's Legacy of Betrayal

Unlike Ames, Philby was motivated by his deep ideological commitment. He was recruited at Cambridge during the Depression, when communism seemed to many intellectuals the only bulwark against fascism.

Philby kept his Soviet allegiance even as he rose to head MI6's Soviet counterintelligence section. The damage Philby inflicted was incalculable. For nearly three decades, he compromised countless Anglo-American operations, resulting in the deaths of many Western agents. Perhaps most devastatingly, he formed a close friendship with James Angleton, who would later become the CIA's counterintelligence chief. Their regular martini lunches in Washington became venues where Angleton unwittingly shared the most sensitive American intelligence operations—information Philby promptly passed to Moscow.

The exposure of Philby's treachery shattered the CIA's foundation of trust. Devastated by the betrayal of his friend, Angleton descended into a paranoid hunt for Soviet moles that paralyzed American intelligence operations for years. This "wilderness of mirrors" that Philby created—where no one could be fully trusted—ironically helped create the environment where Ames could later operate undetected.

The psychological contrast between these two master spies reveals a great deal about the evolution of espionage in the 20th century. Philby represented the ideological spy—someone who believed genuinely in the cause for which he betrayed his country. His Cambridge contemporaries—Guy Burgess, Donald Maclean, and Anthony Blunt—formed a spy ring motivated by similar political convictions.

Ames, operating in the more cynical 1980s, represented something more mercenary—betrayal as a simple transaction. This shift from ideology to financial motivation would become increasingly common in later espionage cases, including those of Robert Hanssen at the FBI and Jonathan Pollard at Naval Intelligence.

Both men's betrayals exposed critical vulnerabilities in Western intelligence. Philby's case revealed the dangers of a system that relied too heavily on social class and personal connections for security vetting—his "right" background and elite education placed him above suspicion for years. Decades later, Ames's case highlighted the failure to monitor financial anomalies—despite spending far beyond his government salary, his extravagance went uninvestigated for nine years.

The Cold War spy world these men inhabited had its own peculiar rituals and codes. For Philby, spying involved midnight meetings in London parks and leaving chalk marks on mailboxes to signal dead drops. Ames, operating in the 1980s, used more sophisticated methods—including encrypted communications and prearranged signal sites around Washington where he would place marks indicating his need for a meeting.

What links these cases across time is the devastating human cost of betrayal. When Philby exposed a CIA-MI6 operation to infiltrate communist Albania in the early 1950s, dozens of agents parachuted behind the Iron Curtain were captured and executed. When Ames revealed Soviet officials working for the CIA, they faced similar fates—and their families often suffered as well.

The legacy of these betrayals continues to mold intelligence operations today. Modern security clearance processes now include sophisticated financial monitoring, psychological profiling, and artificial intelligence systems that flag behavioral anomalies. Yet the perfect system is elusive—the mind determined to deceive will find ways to do so.

Perhaps the most unsettling lesson from these historic betrayals is how ordinary they can appear from the outside. Both Ames and Philby kept convincing facades of normality—attending office parties, mentoring younger colleagues, and partaking in the daily rituals of intelligence work—while simultaneously orchestrating devastation from within.

In the end, these master spies remind us that history's great betrayals don't require exceptional people—just ordinary people making devastating choices in secret. One compromised document at a time.

Psychological Profile: The Making of a Traitor and the Walker Family Spy Ring Structure

After looking at cold-blooded traitors like Aldrich Ames and Kim Philby, let's look at something even more disturbing—a spy who turned betrayal into a family business.

John Walker Jr. didn't just betray his country; he recruited his own flesh and blood for treason. The former Navy communications specialist created what the FBI called "the most damaging spy network in American history" by pulling his son Michael, brother Arthur, and best friend Jerry Whitworth into his operation.

Most spies act alone. Walker built a criminal family enterprise. He weaponized family loyalty.

What started Walker down this dark path? Plain old money problems. Drowning in debt and alimony payments after a bitter divorce in 1967, he walked into the Soviet Embassy in Washington with a simple offer: classified Navy documents in exchange for cash.

The information he provided was gold. As a communications specialist with access to cryptographic keys, Walker gave the Soviets the ability to decode Navy messages for nearly two decades. Soviet submarines suddenly knew exactly where American carriers were positioned. But Walker's most sinister move came when he left active duty. Needing new sources of classified material, he manipulated his own son Michael into joining the Navy to steal documents.

Other fathers took their sons fishing. Michael's father took him into espionage.

Walker's tactics with family ran on pure manipulation. He convinced his brother Arthur, a Navy officer struggling with bills, that spying was easy money. When his daughter Barbara suspected his activities, he threatened her life. The psychological control was complete—he transformed family love into a weapon.

This family-based method posed unique dangers. Security investigators typically look for outside influences that might compromise clearance holders, rarely suspecting that a father might be pushing his son to photograph classified documents during holiday visits.

The house of cards finally collapsed in 1985 when Walker's ex-wife Barbara, after years of abuse, called the FBI. The damage assessment stunned investigators. No matter what Walker gave the Soviets, they had to assume they had everything.

What makes the Walker case so intriguing is how it merges family psychology with espionage. Unlike ideological spies motivated by political belief, Walker was driven by something baser—a need for money coupled with the arrogance to believe he could outsmart everyone.

The case exposed serious blind spots in security protocols. Walker's betrayal began with ordinary pressures many people face—mounting bills, failed relationships, and career frustration. These common stresses created an opening that Soviet intelligence masterfully exploited.

Soviet handlers played Walker perfectly, offering not just money but the respect and recognition he felt the Navy denied him. They treated him as special, feeding his ego while emptying his conscience.

The human cost was devastating. Walker died in prison in 2014. Michael, Arthur, and Jerry all served lengthy sentences. His family remains fractured to this day—some never recovering from the betrayal by the man who should have protected them.

Perhaps the most chilling part isn't the elaborate scheme, but how ordinary it all was. Walker didn't need high-tech gadgets or master disguises. His most effective tool was simply telling his son to bring those documents home for the family.

The Walker case proves that sometimes the most dangerous enemy isn't the foreign agent. Instead, it's the trusted father sitting across the dinner table, asking for just one small favor.

Intelligence That Changed History

The photograph on President Kennedy's desk showed what appeared to be construction equipment in Cuban fields. To most eyes, these might have appeared as harmless agricultural structures. But to the trained analysts at the CIA's National Photographic Interpretation Center, they revealed something terrifying. Soviet SS-4 nuclear missiles were being installed just 90 miles from Florida.

Few stories rival the drama of those thirteen days when humanity stood at the nuclear brink. But what's often overlooked is how intelligence—gathered by ordinary men and women working in extraordinary circumstances—provided the edge that helped pull the world back from catastrophe.

These stories reveal how behind-the-scenes intelligence work quietly influenced three of history's most defining moments: the Cuban Missile Crisis, the secret war in Afghanistan, and the brilliant technological sabotage campaign that helped bring down the Soviet Union. Each represents a masterclass in how espionage, when perfectly executed, can accomplish what armies cannot.

"They've Got Missiles in Cuba"

Richard Heyser wasn't trying to be a hero when he climbed into his U-2 spy plane before sunrise on October 14th, 1962. The Air Force major was simply following orders to photograph western Cuba. But the 928 pictures he brought back would change the course of history.

President Kennedy was skeptical. He most likely looked at those U-2 photos and asked, "Are you sure these are missiles?" Analysts would have to walk him through exactly what they were seeing—missile transporters, erectors, launch pads—all in various stages of construction.

These images gave Kennedy something invaluable: proof that Soviet Premier Nikita Khrushchev lied when he denied placing offensive weapons in Cuba. But gathering intelligence during the crisis extended well beyond that first discovery.

As tensions escalated, Navy RF-8A Crusader jets began daily low-altitude surveillance missions nicknamed "Blue Moon." Flying just 500 feet above Cuban soil at nearly 600 mph, these pilots risked everything to bring back detailed imagery.

The intelligence provided by these missions proved to be invaluable in three ways.

First, it gave Kennedy irrefutable evidence to show the world, including skeptical allies like British Prime Minister Harold Macmillan, who initially questioned American claims about the missiles.

Second, it allowed military planners to identify exactly which sites to target if an airstrike became necessary and just how many Soviet personnel would be killed—an important factor in assessing how the USSR might respond.

Third, and perhaps most importantly, intelligence revealed that while the missiles were being assembled rapidly, they weren't yet operational, buying precious time for diplomacy.

This intelligence victory hinged on human courage as much as technology. A Cuban asset codenamed TOUCHDOWN, later revealed to be Oleg Penkovsky, risked execution to smuggle out details about missile sites. U-2 pilot Rudolf Anderson paid the ultimate price when his aircraft was shot down over Cuba on October 27th, becoming the crisis's sole combat fatality.

Khrushchev had gambled that he could secretly deploy the missiles before America discovered them. When Kennedy proved that he knew exactly what was happening, the psychological impact of being caught in a lie forced the Soviet leader to recalculate. Intelligence had turned the tables on deception.

How the CIA Broke the Soviet Military in Afghanistan

A world away from Cuba, in the rugged mountains of Afghanistan, another intelligence triumph was unfolding two decades later. This triumph would help trigger the collapse of the Soviet empire itself.

By 1985, the Soviet occupation of Afghanistan had descended into a brutal stalemate. Soviet helicopter gunships dominated the skies, particularly the fearsome Mi-24 Hind—so heavily armored it could withstand small arms fire while raining down destruction on Afghan resistance fighters.

The Afghans called the Hinds "Satan's Chariots." When they appeared, there was nothing to do but hide and pray.

The CIA's solution to this aerial dominance would transform the war: FIM-92 Stinger anti-aircraft missiles.

Getting these advanced weapons to the mujahideen presented extraordinary challenges. Each missile cost $38,000 (about $93,000 today), weighed 35 pounds, and required specialized training. The entire operation had to remain covert, as official American involvement would risk direct confrontation with the Soviet Union.

The breakthrough came through an unlikely partnership between CIA officer Michael Vickers (later Assistant Secretary of Defense), Texas Congressman Charlie Wilson, and Pakistani intelligence chief General Akhtar Abdur Rahman. Together, they created a pipeline to smuggle hundreds of Stingers across the Pakistani border and train Afghan fighters to use them—all while maintaining plausible deniability.

The first successful Stinger strike came on September 25th, 1986, when Afghan fighters shot down three Soviet helicopters near Jalalabad. Word spread rapidly through mujahideen channels, and the psychological effect was immediate.

The Soviets completely changed their tactics overnight. Helicopter pilots refused to fly below 12,000 feet—too high for effective ground attacks. Fighter-bombers began dropping bombs from altitudes that made pinpoint accuracy impossible. They didn't just give the Afghans a weapon—they broke Soviet air power.

Within months, the Stingers had achieved near-mythical status among Afghan fighters. Soviet pilots began reporting "Stinger panic"—a fear so paralyzing that some refused missions entirely. By 1987, Soviet air losses had become unsustainable, with over 270 aircraft downed.

But the Stinger's greatest impact was psychological. The Stingers shattered the aura of Soviet military invincibility. Young Soviet conscripts realized they were fighting a war their leaders couldn't win. That psychological blow resonated all the way back to Moscow.

Mikhail Gorbachev later called Afghanistan "the bleeding wound" that drained Soviet resources and morale. By February 1989, the last Soviet troops withdrew from Afghanistan—a humiliating defeat that accelerated the unraveling of the Soviet empire itself.

The operation succeeded because it targeted a critical vulnerability in Soviet military psychology. Soviet doctrine emphasized technological superiority and overwhelming firepower. When a simple shoulder-fired missile neutralized their advanced aircraft, the entire doctrinal foundation began to crumble.

The human toll was immense on all sides. Over 15,000 Soviet soldiers died in Afghanistan. Hundreds of thousands of Afghans perished. And the operation came with unforeseen consequences—many fighters trained and armed by the CIA later joined the Taliban or Al-Qaeda.

Intelligence victories often solve immediate problems, but they also often create future ones. America was focused on defeating the Soviets, not on what Afghanistan might become afterward.

The Greatest Act of Technological Sabotage in History

While Afghan fighters were downing Soviet helicopters with American missiles, another intelligence operation was targeting something even more monstrous—the Soviet technological base itself. This operation would combine deception, psychological warfare, and technological sabotage in perhaps the most creative covert action of the Cold War.

It began with a stroke of luck. In 1981, French intelligence recruited a KGB officer code-named "Farewell"—Colonel Vladimir Vetrov, who supervised the KGB's scientific and technological espionage program called Line X. Disillusioned with Soviet corruption and drawn to Western freedoms, Vetrov handed over thousands of documents revealing how the Soviets systematically stole Western technology rather than developing their own.

What Vetrov gave the U.S. was astonishing. It wasn't just that they were stealing U.S. technology—it was that their entire industrial base had become dependent on theft rather than innovation.

When President Reagan shared this intelligence windfall with CIA Director William Casey, they devised an audacious counter-operation: instead of simply plugging the security leaks, why not feed the Soviets modified technology designed to fail?

The CIA worked with American and European companies to plant flawed computer chips, sabotage turbine designs, and implement software loaded with hidden "time bombs" in the way of Soviet industrial spies. The KGB eagerly collected these poisoned "secrets" and brought them home for implementation in Soviet military and civilian systems.

The results were spectacular. In 1982, a Soviet gas pipeline in Siberia exploded with the force of a small nuclear weapon. This disaster was caused by sabotaged control software that deliberately increased pressure, leading to the pipes bursting. The explosion was so massive that it was visible from space and initially mistaken by some analysts as a nuclear test.

Across the Soviet Union, manufacturing plants installed with stolen Western technology began producing defective goods. Military systems failed during testing. Computer networks crashed mysteriously. Diesel engines on naval vessels stopped working mid-voyage.

This psychological warfare against Soviet industry was effective because it exploited a fundamental weakness in the Soviet system: its increasing dependence on stolen rather than indigenous technology. The operation forced the Soviets to waste billions on parallel research programs to replace technology they no longer trusted, accelerating their economic decline.

The human factor was again decisive. Vetrov betrayed his country partly due to ideology but also because of personal grievances—he felt undervalued and passed over for promotion. The Soviets eventually caught and executed him, but not before he'd delivered over 4,000 documents that provided the blueprint for their technological unraveling.

Why Intelligence Changes History

These three operations—Cuba, Afghanistan, and the Farewell Dossier—share common characteristics that explain their outsized historical impact.

First, they targeted psychological vulnerabilities, not just physical ones. In Cuba, Khrushchev's deception backfired catastrophically when exposed. In Afghanistan, Soviet pilots' fear undermined their technological advantage. With the Farewell operation, Soviet industrial pride turned to doubt and paranoia when technology failed.

Second, they were executed by people who deeply understood their adversaries. The photo analysts who spotted the Cuban missiles had studied Soviet military equipment for years. The CIA officers who implemented the Stinger program knew exactly how Soviet air tactics worked and how to counter them. The architects of the Farewell operation understood the Soviet technological dependency on theft.

Third, they combined technical means with human sources. U-2 cameras photographed missile sites in Cuba, but ground agents confirmed which ones had become operational. Stinger missiles destroyed helicopters, but human networks distributed them to the right fighters at the right moments. Technical sabotage damaged Soviet systems, but a human spy knew exactly what to target.

The ripples from these operations continue today. The photographic techniques developed for Cuba evolved into satellite systems that now monitor everything from North Korean missile sites to climate change. The lessons from Afghanistan contributed to today's counterinsurgency doctrine. And the cyber sabotage techniques pioneered in the Farewell operation have evolved into today's digital battlefield tactics, most dramatically seen in operations like Stuxnet against Iran's nuclear program.

Intelligence is history's invisible hand. The public sees the results—a crisis averted, a war ended, an empire fallen—but rarely the careful work that made it possible.

Reflections

The damage done by a single double agent can eclipse the work of a hundred loyal operatives. Aldrich Ames betrayed CIA assets with a casual signature that sent at least ten American sources to their execution. Kim Philby's cocktail party charm masked his systematic dismantling of Western intelligence operations for nearly three decades. Their betrayals weren't just professional failures—they were death sentences for those who trusted them.

What drives someone to betray? Money motivated Ames, who sold secrets to fund his taste for Colombian neckties and Jaguar sedans. Philby's ideological devotion to Soviet communism proved more seductive than loyalty to colleagues who considered him family. Robert Hanssen, the devout Catholic FBI agent who spied for Russia, sought validation that his bureau career never provided.

The most chilling lesson is that these men weren't obvious villains. They attended the same Christmas parties as their victims. They coached Little League, sang hymns in church pews, and passed the most rigorous security screenings. Their betrayals blindsided institutions because they perfected the appearance of trustworthiness.

Intelligence agencies have learned painful lessons from each case. Polygraphs and background checks won't catch a spy who believes their own cover story. Unexplained wealth should trigger immediate investigation, not office gossip. Compartmentalization of information isn't paranoia—it's prudence.

History shows that espionage's greatest vulnerability isn't technical—it's human. No encryption can protect against the insider who already has the key. No counterintelligence program can fully shield against the colleague with twenty years of sterling service who makes a single fateful decision.

These betrayals force us to confront uncomfortable truths about human nature while reinforcing why intelligence work matters. For every traitor whose name we know, countless loyal officers have lived and died in anonymity, protecting principles larger than themselves.

Intelligence communities face the challenge of building trust while acknowledging their limits. The balance isn't perfect, but the stakes— measured in lives lost when trust is misplaced—demand nothing less.

Chapter 10:

Digital Shadows—Espionage in the Information Age

The spy who changed history no longer looks like James Bond or Mata Hari. They are a 28-year-old man in Shanghai who never leaves his desk, or a former female math major in a nondescript Maryland office park. Her weapon isn't a silenced pistol but a carefully crafted line of code that can cripple a nuclear facility, steal an election, or turn your refrigerator into a listening device.

This secret war operates by strange rules. America publicly condemns China for stealing intellectual property while the NSA taps underwater cables to vacuum up global communications. Russia denies election interference while the country's military intelligence creates increasingly intricate phishing emails targeting political campaigns worldwide (Fessler, 2019).

The lines between crime, espionage, and warfare have dissolved. The same North Korean hackers who stole $81 million from Bangladesh's central bank also unleashed the WannaCry ransomware that shut down hospitals across Britain. Was this a bank robbery, an intelligence operation, or an act of war?

Perhaps the most disturbing part was that, unlike traditional espionage, these digital weapons can be stolen and repurposed. Criminal groups have exploited leaked NSA hacking tools to seize control of systems worldwide.

When Edward Snowden revealed the extent of America's global surveillance in 2013, he exposed a digital dragnet beyond anything George Orwell imagined—a system capable of tracking billions of communications daily.

The spy war surrounding you right now may be inconspicuous, but its consequences couldn't be more real.

The Cyber Landscape of Espionage

The CIA officer stood anxiously beside Admiral Bobby Ray Inman, Deputy Director of the Agency. On the screen before them played real-time satellite imagery of a massive explosion ripping through the Siberian wilderness.

"Did we just start a war?" the officer might have muttered.

"No," Inman replied with a thin smile. "We just ended one."

The blast was no accident. It was the result of Operation FAREWELL, a CIA plan to feed defective computer code to Soviet spies. The code, embedded in pipeline control software stolen by KGB agents, had just caused a three-kiloton explosion in a Soviet gas pipeline—all without a single American agent on Russian soil.

This 1982 operation marked the birth of cyber warfare—a revolution in espionage that transformed how nations attack, defend, and spy on each other.

The evolution from Cold War spy craft to today's digital operations represents the greatest transformation in intelligence since the development of signals intelligence in World War II.

Let's look at how cyber capabilities have created entirely new forms of espionage—and how three major case studies reveal what this means for our future.

Stuxnet

In June 2010, a security researcher investigated a strange computer worm found in systems across Iran. What he found left him stunned. It was the most complex malware ever created, specifically designed to physically destroy Iran's nuclear centrifuges while hiding its presence from operators.

The worm, later named Stuxnet, represented a quantum leap in cyber capabilities. While previous malware stole information or crashed computers, Stuxnet did something unprecedented. It attacked physical infrastructure by taking control of the Siemens industrial controllers operating Iran's uranium enrichment centrifuges (ET Online, 2024).

After infiltrating Iran's nuclear facility at Natanz through infected USB drives (Iran's systems weren't connected to the internet), Stuxnet hunted specifically for the exact configuration of centrifuges used in uranium enrichment. When it found them, it secretly recorded normal operations and then played these recordings back to monitoring systems while simultaneously commanding the centrifuges to spin at damaging speeds.

The Iranians couldn't understand what was happening. While monitoring systems indicated everything was fine, their centrifuges were being destroyed.

The damage was severe. Nearly 1,000 IR-1 centrifuges were destroyed, and Iran's nuclear program was delayed by an estimated 18 months to two years (*Nuclear Power in Iran*, 2024). No bombs were dropped, no commandos inserted, no international crisis triggered—yet the mission was successful.

Though never officially acknowledged, Stuxnet is widely attributed to a joint American-Israeli operation codenamed "Olympic Games." It began under President George W. Bush and accelerated under President Obama. This project represents the first time a cyberattack was used instead of conventional military options to accomplish a critical national security objective.

The operation required extraordinary intelligence work beforehand. Operatives needed detailed knowledge of Iranian nuclear facilities, the exact models of centrifuges, and even the specific Siemens controllers being used. This information likely came through traditional human intelligence in conjunction with earlier cyber intrusions that mapped Iran's systems.

Stuxnet marks a pivotal moment when digital operations moved from theory to reality. Just as the atomic bomb transformed warfare in 1945, Stuxnet proved how code could achieve objectives previously requiring kinetic military action.

Operation Cloud Hopper

While Stuxnet demonstrated how cyber capabilities could cause destruction, another massive operation revealed how they could steal on a massive scale.

In 2016, computer security firms uncovered evidence of what they called "one of the largest sustained global cyber espionage campaigns." Nicknamed "Cloud Hopper," the operation targeted managed IT service providers—companies that handle tech operations for thousands of businesses worldwide.

It was breathtakingly ingenious. Instead of trying to break into thousands of companies individually, they compromised the IT providers who had legitimate access to their clients' networks.

The attackers, later identified as a Chinese state-sponsored group called APT10 (Advanced Persistent Threat 10), gained access to IT service providers like IBM and HPE, then used that foothold to infiltrate these companies' clients—effectively "hopping" from the cloud providers to their targets (*Chinese State-Led "Cloud Hopper" Campaign*, 2024).

The scale was staggering. Through these initial breaches, Chinese operatives gained access to hundreds of companies across 12 countries. These companies included some of the world's largest firms in banking, telecommunications, consumer electronics, medical equipment, pharmaceutical manufacturing, and oil and gas production.

They didn't randomly choose targets. The operation methodically focused on companies with valuable intellectual property, advanced manufacturing techniques, or strategic data. Once inside, they extracted hundreds of gigabytes of industrial secrets and proprietary technologies.

This mission wasn't random cybercrime. It was state-sponsored economic espionage on an industrial scale.

In 2018, the U.S. Department of Justice indicted two Chinese nationals affiliated with APT10. The indictment claimed they had stolen "hundreds of gigabytes of sensitive data" from more than 45 U.S. technology companies and government agencies (*Two Chinese Hackers*, 2018).

Cloud Hopper is historically significant because it illustrates the evolution of intelligence gathering from individual documents to wholesale digital theft. While Cold War spies might spend years grooming an agent to access a single file cabinet, digital operations can extract entire corporate databases in minutes.

The economics of espionage have completely changed. For the investment of a few skilled hackers, nations can steal billions in research and development.

The operation also exposed a new vulnerability of the digital age: the interconnectedness of systems. Companies that believed their security was robust realized they were only as secure as their weakest service provider.

PRISM and XKeyscore

In June 2013, *The Guardian* published a story that would change how citizens worldwide viewed digital privacy. Based on documents leaked by former NSA contractor Edward Snowden, it was revealed the existence of PRISM, a massive NSA surveillance program collecting data directly from the servers of major internet companies, including Google, Facebook, Microsoft, and Apple (Greenwald & MacAskill, 2013).

The leaked documents revealed not just PRISM but an entire ecosystem of surveillance programs. One particularly powerful system called XKeyscore allowed analysts to search through billions of records using just a person's name, email, IP address, or other identifiers.

While U.S. officials disputed some of Snowden's characterizations, the documents confirmed the massive scale of digital intelligence gathering. Unlike traditional targeted surveillance, these programs collected large amounts of data from millions of people, then used algorithms and search tools to find relevant information.

The intelligence community defended these programs as essential for counterterrorism. They credited PRISM with helping to prevent numerous terrorist plots, including a 2009 plan to bomb the New York subway system.

The Snowden revelations marked a watershed moment, during which the public gained insight into modern signals intelligence capabilities. They revealed that in the digital age, the limiting factor in intelligence gathering is no longer collection but analysis. Agencies can access more data than they can meaningfully process.

The programs also showed how modern espionage has inverted the traditional intelligence model. Rather than starting with suspects and getting information about them, systems collect data on everyone, then apply tools to identify patterns and persons of interest.

It's like storing the entire haystack in hopes you'll need a needle later.

The revelations triggered intense international backlash, strained diplomatic relations, and prompted significant reforms to U.S. surveillance laws. The 2015 USA FREEDOM Act imposed stricter limitations on bulk data collection, although many capabilities remained intact.

Social Media and Digital Manipulation

In 2016, Lisa Harris noticed something odd about the Black Lives Matter rally she'd been invited to on Facebook. The event page looked legitimate, with passionate posts about racial justice and hundreds of fellow Floridians planning to attend. She shared it with friends and prepared protest signs.

What Harris didn't know was that she'd been targeted by operatives working for Russian military intelligence. The rally wasn't organized by American activists but by agents sitting in an office building in St. Petersburg, Russia. The "Americans" she interacted with online were fabricated personas managed by Russian military officers.

It's the biggest change in the craft since radio came along. Social media has created an entirely new battlefield where intelligence agencies wage wars for influence, information, and psychological advantage.

For the first time in history, intelligence agencies can interact directly with foreign populations on a massive scale, without going through government or media intermediaries.

How did platforms designed to connect friends become weapons of state power? And what does this mean for the future of espionage?

The Perfect Intelligence Machine

From a spy agency's perspective, social media platforms offer extraordinary advantages that previous generations of intelligence officers could only dream of.

For starters, they provide unprecedented access to personal information. Users voluntarily share details about their political beliefs, family connections, work history, daily movements, and personal relationships—information that Cold War spies risked their lives to collect.

In addition, social platforms allow intelligence agencies to reach millions of people directly with precisely tailored messages. While Cold War propaganda required expensive radio stations or newspaper plants, today's influence operations can create authentic-seeming "grassroots" movements for a fraction of the cost.

Also, these platforms offer detailed feedback. Intelligence officers can test messages, see which ones resonate, and adjust strategies in real-time based on engagement metrics like shares, comments, and click-through rates.

Social media allows operatives to hide in plain sight. The distinction between a genuine grassroots political movement and an influence operation becomes dangerously impossible for users to discern.

Russia's Social Media Handbook

No country has used social media for intelligence purposes more effectively than Russia. Other nations now emulate its pioneering playbook.

The Internet Research Agency (IRA), established in St. Petersburg around 2013, developed these techniques. By 2016, it employed hundreds of people working 12-hour shifts to create and manage thousands of fake American social media accounts across platforms like Facebook, Twitter, Instagram, and YouTube.

The 2016 Russian operation focused on exploiting existing social divisions rather than creating new ones. They operated accounts pretending to be Black Lives Matter supporters and others posing as Blue Lives Matter advocates (*Committee Sensitive—Russia Investigation*, n.d.). They created pages for Texas secessionists and Muslim activist groups. Their goal wasn't to promote a specific ideology but to amplify the most extreme voices on all sides.

What made these operations a success wasn't just their scale but their localization. Russian operatives studied American speech patterns, cultural references, and regional issues. They referenced local news stories and created events in specific cities, giving their fake personas credibility.

They understood that people trust their neighbors more than the government or the media. By pretending to be fellow Americans concerned about local issues, they gained an authenticity that traditional propaganda could never achieve.

The operation wasn't limited to posts and memes. The IRA organized at least 129 real-world events through their fake social media accounts, successfully mobilizing Americans to attend rallies, protests, and demonstrations—sometimes orchestrating opposing groups to confront each other at the same location.

One particularly striking example occurred in May 2016, when the IRA organized both a "Stop Islamization of Texas" protest and a "Save Islamic Knowledge" counter-protest at the same time and location in Houston (Hendrix, 2021). Americans on both sides showed up to protest, unaware that both events had been created by the same Russian operators.

Digital Deception and False Flags

Beyond creating fake grassroots movements, intelligence agencies have developed techniques to spread misinformation and confusion through social media.

One powerful tactic is the "false flag" operation—actions designed to appear as though another country or group carried them out. In the digital age, this involves creating cyberattacks or influence campaigns that mimic another nation's "digital fingerprints."

In 2019, Reuters reported that the U.S. Cyber Command had carried out operations against Iran that were designed to look like they came from other countries. Similarly, Russian hackers have been observed planting false flags in their code, including Russian-language comments deliberately left as red herrings.

These false flag operations create strategic ambiguity that paralyzes action. If a country can't be certain who attacked it, responding becomes politically difficult.

Another highly effective strategy is the "leak and amplify" tactic—where intelligence agencies hack sensitive materials, selectively leak them, then use social media to gain maximum exposure and impact.

The most notorious example took place during the 2016 U.S. presidential election, when Russian military intelligence (GRU) hacked the Democratic National Committee and Clinton campaign chairman John Podesta. They then leaked stolen emails through WikiLeaks and other outlets. Russian-controlled social media accounts and bots then amplified this content, making sure it reached the right audiences.

What made this operation a success from an intelligence perspective was how it laundered stolen information through seemingly legitimate channels. Most Americans who read about the leaked emails had no idea they were seeing the fruits of a Russian intelligence operation.

It creates a difficult dilemma for media organizations. The information is newsworthy and authentic, even if obtained illegally through foreign intelligence operations. By using legitimate media as amplifiers, intelligence agencies can give stolen information credibility it wouldn't have otherwise.

Corporate Secrets and Digital Spies

For businesses guarding trade secrets and intellectual property, the social media era has created vulnerabilities. Instead of breaking into offices to photograph documents, today's corporate spies target employees through their online profiles.

LinkedIn has become the primary hunting ground for intelligence services recruiting potential assets. They identify employees with access to valuable information, then approach them under false pretenses—often posing as recruiters or industry peers.

Chinese intelligence services have mastered this method, creating fake profiles that connect with targets for months or years before recruiting them. These operations often begin with harmless-seeming requests—reviewing a conference paper or providing background on publicly available information—before gradually escalating to requests for confidential data.

One technique, called "water-holing," involves compromising websites frequented by employees of target companies, then using these sites to deliver malware. Intelligence agencies study social media to identify which industry blogs, forums, and websites their targets visit regularly.

They don't need to hack your company directly if they can hack the coffee shop website where your engineers hang out online. Social media gives them a map of exactly where to place these traps.

Even casual social media use can reveal damaging information. When employees post about projects they're working on, travel plans, or company events, they're creating intelligence that foreign services can exploit. A seemingly innocent post about visiting a client can reveal business relationships and potential acquisition targets.

The problem is that people compartmentalize their social media presence from their professional security obligations. They don't realize that bits of information across multiple posts can be assembled into valuable intelligence.

The Rise of Non-State Actors

One of the most notable developments in social media espionage is how it has empowered non-state actors to execute operations once limited to governments.

Groups like ISIS pioneered social media recruitment campaigns that borrowed techniques from both marketing and intelligence services. They created tailored content for different audience segments, used data analytics to measure effectiveness, and used encrypted messaging for personalized follow-up—all tactics previously associated with state intelligence agencies.

Hacktivist collectives like Anonymous have launched intelligence-gathering operations and influence campaigns that rival those of nation-states. During conflicts like the Syrian civil war, these groups obtained and publicized intelligence about military movements, government corruption, and human rights abuses.

Even individual influencers with large followings can now run operations resembling traditional intelligence influence campaigns. In 2020, a network of teenagers and young adults on TikTok and K-pop fan accounts organized to register for tickets to a presidential rally in Tulsa, Oklahoma. They had no intention of attending, creating false expectations about turnout that led to an embarrassingly empty venue.

This democratization of abilities once limited to powerful states creates new challenges for traditional intelligence services. During the Cold War, agents knew exactly who their adversaries were. Today, a teenager with a smartphone might be running an influence operation from their bedroom that impacts national security.

When Digital and Physical Operations Merge

Combining social media tactics with traditional espionage creates powerful synergies between digital and physical operations.

The 2018 poisoning of former Russian intelligence officer Sergei Skripal in Salisbury, England, exemplifies this. After British authorities accused Russian military intelligence of the attack, Russia launched a massive disinformation campaign across social media platforms. It promoted at least 138 contradictory narratives about what had happened.

The goal wasn't to convince people of any single alternative explanation. It was to flood the zone with so many competing stories that people would throw up their hands and say, "We'll never know what really happened."

Researchers call this the "firehose of falsehood" technique, where the objective isn't to create a single counter-narrative but to create confusion and doubt through sheer volume of contradictory information.

Similarly, during the COVID-19 pandemic, Russian and Chinese intelligence services spread conflicting conspiracies about the virus's origins—sometimes promoting theories that it was an American bioweapon, other times suggesting it was created in a Chinese lab. The goal was to create confusion, make people doubt everything, and create an environment where all sources seem equally untrustworthy.

It's an advanced psychological operation (PSYOP) strategy that intelligence agencies have refined for the social media era, where algorithms reward emotional engagement over truthfulness.

Defending Against the Invisible Hand

As social media intelligence operations have grown more advanced, governments and platforms have struggled to implement effective countermeasures.

Following the 2016 election interference, Facebook, Twitter (now X), and other major platforms introduced new policies and detection systems specifically targeting coordinated inauthentic behavior. Their goal is to stop accounts and pages that work together to mislead people about who they are and what they're doing.

These efforts have had some success. Between 2017 and 2021, Facebook removed over 150 networks engaged in coordinated inauthentic behavior. These networks included operations from Russia, Iran, China, and even domestic political groups. X has similarly dismantled thousands of state-linked information operations.

Government responses have also evolved. The U.S. Cyber Command now conducts "hunt forward" operations, working with allied nations to identify and counter foreign influence campaigns before they can target American audiences. The European Union set up the East StratCom Task Force specifically to detect and respond to Russian disinformation campaigns.

Perhaps most importantly, civil society organizations and independent researchers have devised techniques to identify influence operations. Groups like Graphika, the Atlantic Council's Digital Forensic Research Lab, and the Stanford Internet Observatory use network analysis, linguistic patterns, and technical indicators to expose covert campaigns.

The Future of Social Intelligence

In the future, a few emerging technologies promise to make social media intelligence even more powerful.

Artificial intelligence, particularly generative AI systems like GPT-4, can now create authentic-seeming text, images, audio, and video at scale. These technologies could allow intelligence agencies to operate thousands of seemingly human accounts with minimal human oversight (*Militarising Big Tech*, 2023).

Deepfake technology—AI-generated video and audio that can realistically mimic real people—creates opportunities for deception. Intelligence agencies could create convincing fake videos of political leaders making inflammatory statements or issuing orders, which could potentially trigger real-world crises.

The fragmentation of social media also creates new opportunities for targeted operations. As users migrate from public platforms like Facebook to encrypted messaging apps like Signal and Telegram, influence operations can become increasingly difficult to detect and counter.

Reflections

The spy game has changed forever.

Moscow's SVR officers still meet assets in parks, but their real prizes come from server breaches that steal millions of personnel files. Chinese intelligence still recruits human sources, but harvests far more from artificial intelligence sifting through billions of digital breadcrumbs we carelessly leave behind. The CIA still trains field operatives, but increasingly deploys keyboards rather than disguises to penetrate adversaries.

This digital battleground has created a disturbing reality: Today's threats evolve at network speed, yet the laws we follow were designed for a world that relied on paper. When a foreign intelligence service steals data from American companies, is that espionage (traditionally accepted) or an act of war? When social media manipulation drives citizens to violence, who is to blame? When malware jumps from its intended target to hospitals and power grids, where do we draw the line?

The uncomfortable truth is that we've built a global digital infrastructure optimized for speed and convenience rather than security. We've created intelligence tools of enormous power, then watched them leak into the hands of criminals and terrorists. We've connected everything—from power plants to pacemakers—to networks that remain fundamentally vulnerable.

What happens next depends on the choices we make today. Will we demand accountability from tech companies that collect more information about us than the East German Stasi ever dreamed possible? Will we insist on international norms governing digital weapons as we once did for chemical and biological agents? Will we find the balance between security capabilities and privacy protections?

History offers a sobering lesson: powerful tools rarely remain unused. The tools built for targeted intelligence gathering inevitably expand to broader applications. The techniques pioneered by major powers eventually spread to smaller nations and non-state actors.

The digital world is still a wild beast without established rules or boundaries. The question is, will it become a space of repression or a space of security and liberty?

Chapter 11:

Spies on Screen and Page—Truth vs. Fiction

The most dangerous thing about James Bond isn't his license to kill—it's how he's hijacked our understanding of intelligence.

While real CIA officers sift through Excel spreadsheets in suburban Virginia, their fictional counterparts bed supermodels in Monaco and engage in elaborately choreographed rooftop chases. When MI6 struggles to justify its budget to Parliament, 007's influence is larger than any actual operation they've undertaken. It's not just entertainment—it's a distortion that affects recruitment, public policy, and the work of intelligence agencies.

The irony runs deeper when you realize that many authors of spy novels were people who worked in real intelligence.

John le Carré (David Cornwell), who spent years in MI6, gave us George Smiley—the anti-Bond whose weapons were his painstaking attention to detail and psychological insight. The haunted protagonists in Graham Greene's books reflect his own troubles in British intelligence (*How John le Carré Reinvented the Spy Novel*, 2020). These writers used fiction to reveal truths about espionage that classified memoirs never could. The moral compromise. The bureaucratic absurdity. The toll on personal relationships.

Their literary counterpoint to Bond wasn't just an artistic choice—it was almost a confession.

Meanwhile, intelligence agencies struggle with their fictional representations. CIA officers report being asked during operations abroad if they "have a license to kill." The agency has a curator-managed relationship with Hollywood, occasionally offering access in exchange for portrayals that serve recruiting goals. After *Zero Dark Thirty* (2012) depicted the hunt for Bin Laden, debate erupted about its accuracy and whether intelligence officials compromised security by cooperating with filmmakers.

Occasionally, fiction gets it right. *Zero Dark Thirty* portrayed the grinding anxiety of deep-cover officers living double lives. *The Lives of Others* (2006) revealed the soul-crushing banality of surveillance.

Slow Horses (2022) showed the unglamorous side of intelligence failures. These stories are correctives to the Bond-ian fantasy, offering glimpses of intelligence work's true nature: methodical, morally challenging, and often maddeningly bureaucratic.

Perhaps most fascinating is how these fictional portrayals loop back to influence the services themselves. When the KGB formed its first specialized assassination unit in the 1960s, they literally named it "Department V"—a nod to the fictional SMERSH. After the success of the TV show *24* (2001–2010), CIA interrogators reportedly emulated Jack Bauer's techniques, despite their questionable effectiveness and legality (Clews, 2014).

Sometimes, even the professionals can't tell where reality ends and fiction begins.

Spies and Tradecraft: Literature Meets Reality

George Smiley sits silently in a dilapidated safehouse, patiently questioning a nervous fugitive. No gunshots rang out in the air. No car chases take place. Words and glances are the only tension. This is John le Carré's vision of espionage—methodical, psychological, morally ambiguous, and often crushingly mundane.

Six time zones away, James Bond orders a martini in a glitzy Monte Carlo casino. Within minutes, he seduces a mysterious woman, takes part in a high-speed chase, and deploys gadgets to escape danger. By evening, he has brilliantly exposed a supervillain's plot for world domination.

These contrasting portrayals of fictional espionage—one literary, one cinematic—have influenced public perceptions of intelligence work for generations. But which comes closer to reality? And why does it matter?

The Spy Who Wrote Novels

When David Cornwell, writing as John le Carré, published *The Spy Who Came in From the Cold* in 1963, he shattered romantic ideas of espionage. Drawing on his experience as an MI5 and MI6 officer, le Carré presented intelligence work as morally compromised, bureaucratically constrained, and psychologically corrosive.

"What the hell do you think spies are?" says Control, the cynical spymaster in the novel. "Moral philosophers measuring everything they do against the word of God or Karl Marx? They're not. They're just a bunch of seedy, squalid bastards like me" (le Carré, 1963).

Le Carré's masterpiece, *Tinker Tailor Soldier Spy* (2011), depicts intelligence work as primarily intellectual—a balancing act of human motivations and organizational politics. His protagonist, George Smiley, spends more time reviewing personnel files and performing careful interviews than participating in physical confrontations. The novel's climactic moment involves not a gunfight but a conversation where the traitor finally reveals himself.

In le Carré's world, the primary skills of a good intelligence officer are patience, psychological understanding, and ethical adaptability. His characters rarely participate in violence themselves. When violence does take place, it's often distant, clinical, and deeply troubling rather than triumphant.

Most distinctively, le Carré portrays espionage as morally ambiguous. In *The Spy Who Came in From the Cold*, the protagonist, Alec Leamas, discovers that his own side is as ruthless and cynical as the enemy. The novel ends not with victory, but with the bitter realization that both the East and the West use and discard individuals in pursuit of strategic advantage.

This moral ambiguity reflects le Carré's own disillusionment with British intelligence, particularly following the 1963 revelation that the "Cambridge Five" spy ring—including high-ranking MI6 officer Kim Philby, introduced in Chapter 8—had betrayed Western secrets to the Soviet Union for decades.

The Quiet Truth of Espionage

If le Carré looked into the institutional and psychological dimensions of espionage, Graham Greene—who worked for MI6 during World War II under the notorious double agent Kim Philby—focused on its human and ethical costs.

Greene's novels, particularly *The Quiet American* (1955) and *The Human Factor* (1978), depict intelligence work not as an adventure but as a form of moral corruption that damages everyone it touches. His protagonists aren't professional spies but ordinary people drawn into espionage by circumstance or misguided idealism.

In *The Human Factor*, protagonist Maurice Castle leaks information to the Soviets not out of ideology but out of personal loyalty—to protect his Black South African wife from apartheid government agents supported by British intelligence. The novel forces readers to confront uncomfortable questions: What do we owe institutions versus individuals? When does personal loyalty trump national loyalty?

Greene highlights the human side of espionage through the setting and local culture in his work. *The Quiet American*, set in 1950s Vietnam, shows how American intelligence officers' ignorance of local culture leads to disastrous consequences. This theme—that effective intelligence work requires deep cultural understanding rather than just technical skills—reflects Greene's own experience in MI6.

Bond, James Bond

When Ian Fleming created James Bond in 1953, he drew on his wartime experience in naval intelligence. Yet Fleming deliberately crafted Bond to be a fantasy—an escapist hero for readers in post-war Britain, when rationing was still in effect and the empire was crumbling.

The film adaptations pushed this fantasy even further, creating a lasting but misleading image of intelligence work. In the Bond films, espionage involves these components:

- constant action and violence, with Bond personally killing dozens of adversaries

- cutting-edge gadgetry that borders on science fiction

- lone-wolf operations with minimal oversight or teamwork

- immediate sexual conquest of beautiful women spotted during missions

- clear moral lines with evil villains

- minimal paperwork, analysis, or patience

Real intelligence officers watch these portrayals with amusement and frustration.

Beyond Bond, Hollywood has created countless takes on the spy thriller, from the hyper-competent Jason Bourne to the comedic Austin Powers.

While more recent films like *Zero Dark Thirty* strive for realism, most cinematic portrayals share common distortions:

- compression of time (operations that would take months or years occur in days)

- exaggeration of individual agency (single operatives accomplish what would require teams)

- over-emphasis on physical skills versus analytical abilities

- simplified moral frameworks with clear heroes and villains

These distortions don't just cause confusion. They also create problematic expectations for the public and potential intelligence officers.

What Real Spies Actually Do

To appreciate how literary and cinematic portrayals diverge from reality, it's worth considering how intelligence officers actually spend their time:

- **Building relationships and networks:** Developing sources who provide information, often over years of patient pursuit.

- **Information gathering and analysis:** Collecting data from multiple sources and analyzing it to identify patterns and significance.

- **Writing reports:** Communicating intelligence findings clearly and accurately to policymakers.

- **Dealing with bureaucracy:** Working within complicated organizational structures with multiple levels of control.

- **Cultural and linguistic immersion:** Understanding the societies where they operate, often requiring in-depth local knowledge.

The skills that make successful intelligence officers rarely resemble James Bond. Instead, they include ones such as these examples:

- patience and persistence

- emotional intelligence and empathy

- analytical thinking and attention to detail

- clear communication skills

- ethical reasoning and judgment

- cultural adaptability

The best intelligence officers aren't the ones who can win a fistfight. They're the ones who can enter a room and make everyone feel comfortable enough to talk openly.

Violence is exceedingly rare in intelligence work. Most officers go entire careers without drawing a weapon. When operations do turn violent, they're typically handled by specialized military units, not case officers whose primary job is human intelligence analysis.

Reality on Screen: Defining True Narratives in Intelligence

The audience sits in awe as Gerd Wiesler, a Stasi agent in 1980s East Germany, carefully adjusts his headphones. From his surveillance position in the attic, the agent monitors a playwright and the playwright's actress girlfriend—both targets of government surveillance. There are no explosions, car chases, or glamorous seductions. Instead, *The Lives of Others* (2006) offers something rarely seen in espionage films: the reality of surveillance work and its corrosive effect on both the watchers and the watched.

The film shows how screen portrayals of espionage can illuminate rather than obscure reality when creators prioritize human truth over spectacle. While Hollywood blockbusters bring entertainment value, it's often films like these that offer viewers authentic glimpses into the intelligence world.

Films That Get it Right

While *The Lives of Others* sets a high standard for authenticity, other films have similarly revealed truths about intelligence work. *Breach* (2007), based on the true story of FBI agent and spy for Russia Robert Hanssen, portrays espionage not as a thrilling adventure but as a psychological chess match played in fluorescent-lit government corridors.

The film reveals that traitors rarely fit into convenient stereotypes. Hanssen wasn't an obvious villain but a complex figure—a devout Catholic family man whose motivations included ego, financial gain, and twisted ideology. This psychological complication reflects the reality of intelligence officers hunting real spies.

Another film praised for its realism is *Tinker Tailor Soldier Spy* (2011), adapted from John le Carré's novel. Unlike action-oriented spy films, it presents intelligence work as primarily intellectual—a systematic process of analyzing data, questioning assumptions, and understanding human psychology.

Small Screen, Big Impact

Television series, with their long-form storytelling, provide some of the most detailed portrayals of intelligence work.

The Americans (2013–2018) follows Soviet deep-cover operatives living as an American family in 1980s Washington, D.C. The series looks at not just their espionage activities but also the psychological strain of maintaining false identities year after year.

The series doesn't shy away from moral conflict, showing protagonists who commit horrific acts for their country while simultaneously experiencing genuine love for their children and questioning their cause. Simple portrayals of heroes and villains don't reflect the reality of intelligence work.

Similarly, *Rubicon* (2010), though short-lived, earned praise for its focus on intelligence analysis rather than field operations. The series features analysts poring over data, finding trends, and making difficult judgments based on incomplete information—the unglamorous but necessary part of intelligence.

The series also accurately depicted the bureaucratic politics of intelligence work—the competition between agencies, the tension between analysis and operations, and the challenge of explaining intelligence to policymakers with predetermined views.

More recent series like *Slow Horses* (2022–present), based on Mick Herron's novels, continue this tradition of authenticity. The show shows disgraced MI5 agents relegated to a department for outcasts, working in decrepit offices rather than high-tech command centers. This acknowledgment of failure as part of intelligence work—that not all officers are elite and not all operations succeed—is a refreshing counterpoint to more sensationalized representations.

When Movies Inspire Real Spy Tech

Throughout spy history, fictional depictions have inspired real operational innovations.

"Q Branch," the fictional technical division in Bond films, became so culturally iconic that it influenced how actual intelligence technical services viewed their missions. Former technical officers acknowledge that creative thinking in fiction can sometimes stimulate similar creativity in reality (Hill, 2024).

Nobody was building ejector seats for Aston Martins, but the general concept—ordinary objects concealing specialized capabilities—absolutely influenced real technical developments. Fiction helped them to think outside the box.

Supposedly, in the 1970s, the CIA's Office of Technical Service created a device similar to the fictional "tracker" seen in the 1964 film *Goldfinger*. The real device used miniaturized technology to follow vehicles—less sophisticated than the fictional version but clearly inspired by the same concept.

Similarly, KGB technical divisions sometimes referenced Western spy fiction in internal documents to propose new stuff (Macdonald et al., 2023). A former KGB technical officer later revealed that American shows like *Mission: Impossible* were screened for technical staff as potential inspiration for countermeasures or adaptable concepts.

When Truth Is Stranger Than Fiction

For filmmakers trying to capture intelligence operations accurately, an intriguing paradox happens—sometimes the truth seems less believable than fiction.

Ben Affleck, who directed *Argo* (2012), a film about the CIA's extraction of American diplomats from Iran using a fake film production cover story, faced this dilemma. The actual operation was even stranger than what was shown. They had to tone down some elements because audiences wouldn't have believed they really happened (Hewitt, n.d.).

All That Glitters, the fake science fiction film created as a cover for the actual CIA operation portrayed in *Argo*, featured details that seemed too convenient to be believable—yet it was based on the actual cover story by CIA disguise specialist Tony Mendez.

It's the same with operational procedures. When Russian agents poisoned Sergei Skripal in Salisbury in 2018, there were so many security failures that a screenplay would have criticized them as unrealistic plot holes.

Reflections

Spy fiction's biggest legacy may be how it infiltrated intelligence agencies, shaping both how the public perceives espionage and how institutions operate.

Consider the curious case of "Q Branch." When Ian Fleming created his fictional MI6 gadget division headed by the enigmatic "Q," no such department existed in British intelligence. Yet by the 1960s, faced with constant public questioning about their version of Q Branch, MI6 set up a technical services division that insiders informally called "Q Branch" (Jasper, 2024). Fiction had literally changed organizational structure.

It's not the gadgets or action sequences that make spy fiction historical, but how it chronicles the changing moral landscape of intelligence work over time. In Graham Greene's 1955 novel, *The Quiet American*, we see the naive blundering of early Cold War operations. Le Carré's *The Spy Who Came In From the Cold* shows the ruthless moral calculus of the 1960s, when idealism gave way to grim pragmatism (Moran &

Hammond, 2021). Post-9/11 thrillers reveal our tortured relationship with surveillance and interrogation techniques that were once unthinkable.

Beyond entertainment, they force us to confront uncomfortable questions: What price are we willing to pay for security? What happens to those who live for years under false identities? How do democratic societies control organizations wrapped in secrecy?

The greatest spy stories remind us that behind every intelligence operation are human beings grappling with impossible choices. Sidney Reilly—the real-life "Ace of Spies" who inspired James Bond—didn't die in a glamorous shootout but was reportedly executed in a Moscow basement after being lured into a Soviet trap. Kim Philby, the most damaging mole in MI6 history, ended his days not dramatically but as a lonely alcoholic in a Moscow apartment.

These human elements—the doubt, the moral compromise, the psychological toll—are what the best spy fiction captures and Hollywood often misses.

With intelligence services playing increasingly visible roles in everything from election security to pandemic response, separating myth from reality becomes not just an academic exercise but a civic duty. Understanding the genuine motivations, limitations, and ethical challenges of intelligence work helps us have more meaningful discussions.

Perhaps that's the final irony: Fiction is our best window into this secretive world, not for its action sequences, but for the human truths beneath the escapism.

Chapter 12:

The Moral Maze—Ethics and

Espionage

Picture an intelligence officer, deeply entrenched in a mission that demands not just skill but profound ethical discernment. In the shadowy confines of their world, decisions are rarely clear-cut. They know they have access to enhanced interrogation techniques, and each decision forces them to ask themselves whether they're staying true to their values or doing what they think must be done. The tension is palpable, particularly as they consider the broader implications of their actions, understanding that each decision could have global consequences.

In this chapter, we delve into these moral quandaries faced by intelligence professionals. We explore how such dilemmas transcend mere protocol to influence not only individual psyches but also international relations, affecting public perceptions and alliances worldwide. Through this exploration, we begin to unravel the profound impact of ethical choices in espionage, setting the stage for further inquiry into the intricate dynamics of intelligence work.

The Necessary Evil?

The dust had barely settled at Ground Zero when America's intelligence began its evolution. The attacks of September 11th, 2001, didn't just collapse buildings—they shattered long-standing boundaries between what was permitted and what was prohibited in the name of national security.

Fast forward a year or so, and the changes driven by this evolution were evident in a cramped interrogation room at a "black site" somewhere in Eastern Europe, circa 2002. An intelligence officer sits across from a suspected al-Qaeda operative. The room is deliberately cold. The prisoner hasn't slept in nearly 48 hours. The officer's hands rest on a folder allegedly containing evidence of the prisoner's guilt, though it might just as well be blank paper—a classic interrogation prop.

He knew what he was supposed to do. His country had been attacked, people were afraid, and his superiors assured him these techniques were both legal and necessary. But something inside him likely kept asking, *Is this who we are now?*

That question haunted countless intelligence professionals during this dark chapter in American history.

The Psychology of Enhanced Interrogation

The post-9/11 program wasn't created from thin air. It drew on psychological research into stress, resistance, and compliance. Two psychologists, James Mitchell and Bruce Jessen, adapted techniques from the military's SERE program (Survival, Evasion, Resistance, Escape). Having completed this training myself in 1996 after earning my Navy Flight School Wings, I can attest to its intensity—the experience is still vivid: freezing, hungry, sleep-deprived, and being pushed beyond our mental limits.

Sleep deprivation pushed detainees to the edge of hallucination. Stress positions caused excruciating pain without leaving marks. Waterboarding triggered primal drowning panic. The CIA paid psychologists $81 million to develop and implement this program (Santhanam, 2014).

Here is what President Bush told the American people in a nationally televised address from the Oval Office on September 11th: (Remarks by the President, 2003).

> The American people need to know that we're facing a different enemy than we have ever faced. This enemy hides in shadows, and has no regard for human life. This is an enemy who preys on innocent and unsuspecting people, then runs for cover. But it won't be able to run for cover forever. This is an enemy that tries to hide. But it won't be able to hide forever. This is an enemy that thinks its harbors are safe. But they won't be safe forever. (Remarks by the President, 2003)

The effectiveness of these methods remains fiercely disputed. Some CIA officials claim they generated valuable intelligence. The Senate Intelligence Committee's exhaustive 6,700-page report concluded otherwise, finding that enhanced interrogation techniques did not yield unique, otherwise unobtainable intelligence (Human Rights First, 2016).

FBI interrogator Ali Soufan, who successfully elicited actionable intelligence from al-Qaeda suspects using traditional rapport-building techniques before the CIA took over with harsher methods, believed that when someone is in pain, they will say anything to get the pain to stop.

The Toll on America's Spies

What is rarely discussed is how these programs corroded the souls of the agents tasked with implementing them. Intelligence officers aren't movie villains—they're patriots who believe they're protecting their country.

They went home every night to their family, tried to be normal, but kept seeing the faces of those men. They told themselves they're the good guys. But how do the good guys balance that with what they were doing each day?

Many developed psychological symptoms indistinguishable from combat-related PTSD: nightmares, emotional numbness, substance abuse, relationship problems (Hersh, 2004). Some requested transfers. Others left intelligence work entirely, unable to align their actions with their self-image.

They faced impossible choices. Should they follow orders they considered wrong or risk their careers by objecting? Many just kept their heads down, but the moral injury was real.

This internal conflict spread throughout the intelligence community. FBI Director Robert Mueller refused to allow his agents to participate in enhanced interrogations, creating tension with CIA counterparts (Reuters Staff, 2008). Military JAG officers raised legal objections. Intelligence analysts questioned whether they could trust information from this source.

Abu Ghraib and the Collapse of Moral Authority

In April 2004, CBS broadcast the first photographs from Abu Ghraib prison, showing American military personnel abusing Iraqi detainees (Leung, 2004). Though military investigations would later distinguish these abuses from officially sanctioned interrogation techniques, the images created a lasting impression in the public consciousness.

The international repercussions were equally severe. British intelligence officers expressed concerns about sharing information collected through torture. German prosecutors investigated CIA officers involved in abductions on German territory. Polish officials faced scrutiny for hosting a CIA black site.

For years, America had lectured these countries about human rights. Now they were throwing back America's own words. The hypocrisy was impossible to defend.

Losing Hearts and Minds

Perhaps the most destructive long-term consequence was losing the trust of local populations whose cooperation was essential for effective counterterrorism.

In 2002, villagers would invite American forces into their homes and share information about Taliban movements. After stories spread about what happened to people they detained, those same villagers would barely make eye contact. Can't blame them—why risk helping people who might throw your brother or cousin into a system like that?

Intelligence collection depends on relationships. Coercion might force short-term compliance, but voluntary cooperation yields far richer information over time. Every person who refused to work with American forces because of fear or moral disgust represented a loss of potentially life-saving intelligence.

The Long Shadow

The debate over enhanced interrogation opened a Pandora's box of ethical questions that continue to plague American intelligence operations. If these techniques could be justified, what else might be permitted in the name of security?

Targeted killing programs expanded dramatically, with drone strikes eliminating terrorist threats but sometimes killing civilians. Mass surveillance programs collected large amounts of data, raising privacy concerns. All of these practices involved difficult trade-offs between security and other values.

Once you start moving ethical boundaries, it gets easier with each step. The question becomes not, "Is this right?" but, "Is this necessary?" And "necessary" is dangerously elastic during a crisis.

The ripple effects reached deep into America's political culture, contributing to increasing partisan division over fundamental values. What began as a debate about specific intelligence techniques evolved into broader questions about American identity itself.

These weren't just policy decisions. They were decisions about who Americans were as a country—whether America's commitment to certain principles is conditional or absolute.

Learning From History

History offers a few interesting moral lessons, especially in espionage. Intelligence work exists in the gray areas between ideals and necessities. Between what we aspire to be and what fear drives us to do.

But this chapter of American intelligence history provides valuable lessons for citizens and future intelligence professionals alike.

First, expedience rarely serves long-term strategic interests. Techniques that undermined America's moral authority damaged intelligence collection capabilities and diplomatic relationships that had taken decades to build.

Second, legal boundaries alone aren't sufficient ethical guidelines. Many enhanced interrogation techniques were blessed by Justice Department memos but failed more fundamental tests of human dignity and American values (Human Rights First, 2016).

Third, institutional culture and leadership matter enormously. When leaders allowed ethical objections, intelligence professionals were more likely to uphold moral boundaries even under pressure.

This period offers a fascinating case study in how democracies respond to threats. The tension between security and liberty isn't unique to post-9/11 America—it has existed throughout history, from ancient Rome to Revolutionary America to the Cold War.

What makes this chapter distinctive is how thoroughly it is documented. The Senate Intelligence Committee report, court cases, memoirs, and declassified documents offer new visibility into decisions once made behind closed doors. Future historians will have rich material to study how fear can transform democratic institutions and values.

As we face new challenges in intelligence and security, the questions raised by enhanced interrogation remain relevant. Where do we draw the line between security and other values? How do we protect our principles under threat? What actions taken in our name are we willing to stand by?

These aren't just questions for intelligence professionals or policymakers—they're questions for every citizen in a democracy. The answers we give will define not just our security but our national character for generations to come.

Targeted Killing Programs and Mass Surveillance Ethics

After Abu Ghraib shocked the world, America's intelligence community didn't retreat from controversial tactics—it simply shifted to less visible ones. As enhanced interrogation techniques faced growing scrutiny, two other programs moved to the center of America's counterterrorism strategy: targeted killings and mass surveillance.

They traded dark rooms for blue skies. Drones don't leave bruises that show up in photos.

The Rise of America's Drone Campaign

Targeted killing is not new—intelligence services have eliminated specific threats for centuries. What dramatically changed after 9/11 was the scale, technology, and legal framework surrounding these operations.

In 2001, the U.S. military had fewer than 50 drones. By 2012, it operated approximately 7,500. The CIA, which historically focused on intelligence gathering rather than paramilitary operations, built its own substantial drone fleet (Rollins, 2023). What began as an exceptional tactic became routine policy.

The appeal was obvious. Drones offered pinpoint strikes without risking American lives. They could reach remote areas where sending ground forces would be politically or militarily impractical. And unlike detention programs, they left no prisoners to interrogate, house, or potentially release. No Guantánamo controversies.

But this tactical convenience came with ethical questions. Unlike traditional battlefield operations, drone strikes often targeted individuals far from active combat zones—in Yemen, Somalia, or Pakistan's tribal areas. The decision to kill rather than capture eliminated any possibility of gathering intelligence from these targets. And the remote nature of drone operations created psychological distance between operators and targets.

Imagine being a drone operator. You see these people living their lives. They play with their kids. Then one day, you're told this person is an imminent threat, and you push the button. Sometimes you watch the aftermath. You see people running to help. You see the grief.

The Al-Awlaki Case

No case better illustrates the ethical dilemma of targeted killing than Anwar al-Awlaki. Born in New Mexico, al-Awlaki was an American citizen who became a prominent al-Qaeda propagandist in Yemen. His online sermons inspired numerous terrorist plots, including the 2009 Fort Hood shooting.

In 2010, the Obama administration placed him on a kill list—the first public acknowledgment that an American citizen had been marked for death without trial. On September 30th, 2011, a CIA drone strike killed al-Awlaki in Yemen. Two weeks later, another strike killed his 16-year-old son, also an American citizen. Officials later claimed the son was not specifically targeted.

The case ignited a heated debate about presidential authority, due process, and constitutional rights. The Fifth Amendment guarantees that no person shall "be deprived of life, liberty, or property, without due process of law." Did a classified legal review within the executive branch satisfy this requirement? Or did it represent an expansion of executive power?

America created a system where the president became prosecutor, judge, jury, and executioner. Even if you trust one president with that power, do you trust every future president?

The al-Awlaki case showed how targeted killing eroded traditional boundaries between war and law enforcement. Between combatants and civilians. Between battlefield and sanctuary.

From Targeting Individuals to Targeting Everyone

As drone strikes targeted specific individuals overseas, a parallel revolution in surveillance capability turned America's intelligence toward mass data collection at home and abroad.

The scope of these programs remained largely hidden until June 2013, when NSA contractor Edward Snowden illegally leaked thousands of classified documents to journalists. The revelations shocked even those familiar with intelligence activities.

Under programs like PRISM, the NSA collected data directly from major internet companies. Through UPSTREAM collection, it intercepted data flowing through the internet's physical infrastructure. BULLRUN involved the deliberate weakening of encryption standards. XKeyscore allowed analysts to search vast databases of emails, online chats, and browsing histories without prior authorization (loek_essers, 2014).

Most controversial was the bulk collection of American telephone records under Section 215 of the USA PATRIOT Act. Every day, telecommunications companies provided the NSA with metadata (who called whom, when, and for how long) on millions of calls made by ordinary Americans—the majority with no connection to terrorism (Senate Refuses to Poison the USA FREEDOM Act, 2015).

General Keith Alexander, who directed the NSA during this period, defended these programs as necessary to "connect the dots" and prevent another 9/11. Critics countered that mass surveillance fundamentally altered the relationship between citizens and government in ways incompatible with democratic values.

The traditional model was that the government could collect information on you only if it had reason to suspect you of wrongdoing. These programs reversed that premise. Everyone became suspicious by default.

The Ethical Costs of Seeing Everything

Both targeted killing and mass surveillance programs sparked ethical questions about the price of security and who pays that cost.

With drone strikes, much of the ethical burden fell on operators. Though physically safe in control rooms thousands of miles from their targets, they experienced unique psychological strains. Many reported symptoms of PTSD despite never being in physical danger themselves.

For surveillance programs, the ethical burden spread more widely through the intelligence community. NSA analysts found themselves with unheard-of access to private communications. Though operating under official guidelines, they faced daily decisions about what was appropriate to collect.

Both programs strained the ethical frameworks of intelligence agencies. Officers trained to believe in American values of due process, privacy, and transparency now operated programs that seemed, to many, at odds with those values.

The Limits of Watchdogs

In theory, thorough monitoring should have prevented ethical overreach in these programs. In practice, the checks and balances weren't enough.

Congressional intelligence committees received classified briefings on these operations but struggled to verify the information they received or assess its completeness. The Foreign Intelligence Surveillance Court approved surveillance activities but heard arguments from only one side—the government. Executive branch reviews happened entirely behind closed doors.

It's worth noting that U.S. Signals Intelligence Directive 18 (USSID 18) outlines the NSA's policies and procedures related to the collection, retention, and dissemination of information about U.S. persons, ensuring compliance with the Fourth Amendment and other legal requirements.

Senator Ron Wyden, who served on the Senate Intelligence Committee, found himself in an impossible position. Bound by classification rules, he could not publicly reveal what he knew about surveillance programs he considered potentially unconstitutional. Instead, he resorted to cryptic warnings, telling a public hearing that Americans would be "stunned" if they knew how government agencies were interpreting surveillance laws.

This regulatory vacuum allowed dubious legal interpretations to go unchallenged. The NSA operated under a legal interpretation that the word "relevant" in Section 215 of the PATRIOT Act could apply to every phone call in America. The Justice Department came up with legal theories justifying the killing of American citizens that remained classified for years.

The problem with secret law is that it's never tested in a court of public opinion. Bad legal reasoning flourishes in the darkness.

Whistleblowers and Public Accountability

When official accountability systems failed, whistleblowers sometimes forced public accountability.

NSA whistleblowers, like Thomas Drake, William Binney, and Kirk Wiebe, tried to raise concerns about surveillance programs through official channels. Most faced retaliation rather than reform. Drake was prosecuted under the Espionage Act, facing potential decades in prison before the case eventually collapsed.

Snowden, seeing these precedents, chose to leak directly to journalists. His disclosures triggered the most significant intelligence reforms in decades, including the USA FREEDOM Act, which ended the bulk collection of American phone records.

Similarly, human rights organizations and journalists exposed details of the targeted killing program that official authorities had kept hidden. The Bureau of Investigative Journalism closely tracked drone strikes and civilian casualties, often contradicting official accounts. Legal challenges from the ACLU and others forced increased transparency about the legal rationales for these operations.

Without these outside forces, the system would never have corrected itself.

The Intelligence Officer's Dilemma

For intelligence officers, these programs created philosophical dilemmas. They joined intelligence agencies to protect their country but found themselves implementing policies that sometimes seemed at odds with the values they thought they were protecting.

Officers involved in the drone program faced equally troubling questions. Decisions about who qualified as an "imminent threat" were based on intelligence that was often fragmentary or ambiguous. Patterns of life used to identify militants sometimes led to targeting errors with fatal consequences. But the pressure to act, to prevent the next attack, was always there.

Crossing Borders, Crossing Lines

When an American drone strike turned Iranian general Qasem Soleimani's car into a fireball on a Baghdad road in January 2020, the smoke had barely cleared before the phones started ringing in intelligence agencies worldwide. European spies shuddered. British officers worried about blowback. Israeli intelligence veterans quietly approved. And in that moment, a truth about modern espionage crystallized: Even allies don't always agree on what's right in the shadows.

This clash of spy ethics wasn't just an academic debate—it had real consequences for intelligence sharing, military cooperation, and diplomatic relations. And it offers us a fascinating window into one of history's most pressing questions: Is there such a thing as a universal code of conduct in espionage, or is it forever destined to be a moral jungle where everyone makes their own rules?

Five Cousins, Different Moral Codes

The most successful intelligence alliance in history began with a handshake. In 1946, American and British signals intelligence leaders formalized their WWII cooperation into what eventually became the Five Eyes—an exclusive espionage club consisting of America, the UK, Canada, Australia, and New Zealand. These English-speaking democracies share almost everything they collect, creating an unparalleled global surveillance network.

You might think seven decades of intimate cooperation would have ironed out their ethical differences. You'd be wrong.

The Americans have always been more aggressive. After 9/11 especially, they pushed boundaries the Brits weren't comfortable crossing. But how do you say no to your most important ally when they're still reeling from the worst attack on their soil since Pearl Harbor?

The differences show up in everything from legal frameworks to operational limits. Australia's intelligence services face tighter restrictions on domestic spying than their American counterparts. Canadian officers need ministerial approval before activities that risk death or detention—a bureaucratic hurdle CIA officers would find maddening. The British have a formal distinction between foreign and domestic intelligence that doesn't match American categories.

These aren't just paperwork differences—they reflect deeper national attitudes about privacy, security, and state power shaped by each country's unique history.

The Aussies remember the Hope Royal Commission in the 1970s that exposed intelligence overreach. The Canadians passed their CSIS Act after their own intelligence scandals. The UK has centuries of balancing an empire's needs against democratic values. These historical experiences created different ethical instincts that persist even as directors come and go.

It can create daily headaches for officers working together. Imagine a Canadian intelligence officer posted to CIA headquarters who gets information through methods her service prohibits. Or an American assigned to Australia's signals intelligence agency who spots a valuable collection opportunity that local laws forbid. The Five Eyes devised elaborate protocols for handling these ethical conflicts, including specialized markings on intelligence products that indicate handling restrictions.

Sometimes, that means agreeing to disagree. Sometimes, it means one service will pass on an operation that the others pursue. What matters is being honest about boundaries.

When Partnerships Get Complicated

If the Five Eyes occasionally stumble over ethical differences despite their shared democratic values, imagine the issues when Western agencies work with partners whose moral compasses point in very different directions.

In Jordan's capital, Amman, a nondescript office building houses the headquarters of the General Intelligence Directorate (GID), one of America's most valuable counterterrorism partners after 9/11. Jordan's spy service provided important intelligence that helped dismantle numerous terrorist networks. They also used interrogation techniques that would land American officers in prison. What do you do?

After 9/11, the answer for many Western agencies was to look the other way—or worse, to exploit these ethical differences through programs like "extraordinary rendition," where terrorism suspects were transferred to countries known to use torture.

The European Court of Human Rights later ruled that Poland, Romania, and Lithuania violated human rights conventions by hosting these CIA facilities. The moral stain spread to partners who tried to claim plausible deniability about what happened on their soil.

Those programs damaged intelligence relationships. They created a trust deficit. Now, European services are much more careful about what they share with American agencies. Always asking, "How will this information be used?'"

Spies, Tech Giants, and the New Moral Gatekeepers

The ethical landscape of international intelligence took another dramatic turn when an unexpected new player entered the scene: Silicon Valley.

The most consequential development in intelligence ethics this century might be the rise of tech companies as moral gatekeepers. Before Snowden, if the NSA had lawful authority to collect information, they could generally get it. Now, they need to convince corporate lawyers that their request meets their ethical standards, not just the government's.

After Edward Snowden's 2013 revelations about mass surveillance, tech companies strengthened encryption and restricted government access to user data. These corporate decisions effectively limited what intelligence services could collect, regardless of what their national laws permitted.

When Apple refused an FBI request to unlock a terrorist's iPhone in 2016, they weren't just making a business decision—they were taking an ethical position about privacy that directly challenged the government's security priorities.

Corporate influence creates surprising ethical tensions. American tech companies may refuse to cooperate with U.S. intelligence on privacy grounds while at the same time accommodating demands from authoritarian governments where they want market access. Chinese companies like Huawei operate under legal obligations to assist Chinese intelligence that far exceed what Western companies face.

The uneven ethical playing field creates real operational disadvantages. Democratic services face constraints from both government regulation and corporate resistance that other adversaries simply don't.

Learning to Move Through the Ethical Maze

How do intelligence officers deal with this moral dilemma? Increasingly, through rigorous ethics training that goes far beyond simple rules.

In a classroom at the CIA's Sherman Kent School, students wrestle with scenarios that would make philosophy professors sweat. What do you do with intelligence from a partner service that might have been gathered through torture? How do you handle a request to share information that could be used for purposes your country prohibits? When do you say no to an ally?

This training increasingly incorporates historical case studies, showing how these dilemmas have evolved over time. Students might examine the Church Committee's 1970s findings about CIA assassination plots. Then trace how those revelations led to executive orders prohibiting political assassinations. Then analyze how those same orders were reinterpreted to allow drone strikes against terrorist leaders.

This isn't just for academics. An intelligence officer in Islamabad who receives information from Pakistan's ISI that might have been coerced needs ethical advice. A signals intelligence analyst in London who finds a partner service targeting his country's citizens needs a plan.

The training helps, but nothing fully prepares you for these moments.

The Friction Points That Won't Go Away

Despite decades of cooperation, certain ethical disagreements trouble even the closest intelligence relationships.

Economic espionage remains a flash point. The United States has long believed that intelligence services should not steal foreign intellectual property to benefit domestic companies—a position that France, Israel, China, and many others have never accepted.

Surveillance targeting allies' citizens creates equally tense situations. The Five Eyes agreement reportedly contains provisions against members spying on each other's populations. However, the Snowden documents suggested these restrictions were sometimes circumvented through creative arrangements.

Even fundamental concepts like "national security" are interpreted differently, creating ethical misalignment. Traditionally, Western democracies separate security threats from economic competition. For China, economic security is inseparable from national security, justifying intelligence collection against foreign companies.

Where Do We Go From Here?

Climate change, pandemic diseases, and other global challenges are changing intelligence priorities—and potentially ethics as well. These kinds of threats require information to be shared across borders, motivating ethical standards to align.

Some encouraging examples already exist. The 2015 agreement between the U.S. and China to prohibit commercial cyber espionage represented a rare instance of an explicit ethical agreement between competing intelligence powers. Though implementation has been patchy, it sets the standard that certain activities should be off-limits, even if they are technically feasible.

Recently, debates between democratic nations about ethical guidelines for AI in intelligence work suggest a growing awareness that new technologies need new ethical guidelines. Shared concerns about authoritarian use of surveillance technologies may drive democratic nations toward more unified moral positions.

These debates connect directly to centuries-old questions about the ethics of state power used in secret. The dilemmas facing today's intelligence officers would be similar to those of Francis Walsingham in Elizabethan England or Wilhelm Stieber in Bismarck's Prussia. The tactics change, but the tension between security and liberty, between national interest and universal values, between operational effectiveness and moral constraints doesn't change.

What makes our current moment unique is the global interconnectedness of both threats and responses. Today's intelligence officers navigate ethical frameworks that cross national, cultural, and technological boundaries in ways that we have never seen before.

As we face an uncertain future filled with new threats and technologies, ethics will determine whether intelligence serves as a force for security and stability. Or, it becomes a tool that undermines the very societies it's meant to protect. History will judge not just whether our intelligence services kept us safe, but whether they preserved the values that made our societies worth protecting in the first place.

Reflections

History has no shortage of intelligence triumphs: the breaking of Nazi Enigma codes that shortened World War II; the critical intelligence that prevented nuclear confrontation during the Cuban Missile Crisis; or the careful cultivation of Oleg Penkovsky, who revealed Soviet weaknesses during the Cold War's most dangerous years.

Yet alongside these victories lies a darker ledger—a record of moral compromises, abandoned agents, and operations that succeeded technically while failing ethically. This history reveals that intelligence services operate not just in physical darkness but in moral twilight.

When Allen Dulles, the longest-serving CIA Director, famously remarked that "intelligence is probably the least understood and most misrepresented of the professions" (Dulles, n.d.), he wasn't just lamenting public misconceptions—he was acknowledging the profession's inherent moral ambiguities. Intelligence work requires asking people to betray their countries, deceive their loved ones, and sometimes sacrifice their lives, all in service to principles that can seem distant and abstract in the moment.

The historical record shows these impossible choices. After World War II, British intelligence recruited Nazi rocket scientists who had used slave labor from concentration camps. During the Cold War, the CIA knowingly worked with war criminals like Klaus Barbie if they provided valuable Soviet intelligence. Following 9/11, debates about enhanced interrogation techniques revealed how quickly democratic societies could rationalize previously unthinkable methods when deeply frightened.

Perhaps the most painful ethical dilemmas involve abandonment. When the CIA trained Hungarian freedom fighters in 1956, it created expectations of American support that never materialized when Soviet tanks rolled into Budapest. During the fall of Saigon in 1975, South Vietnamese agents who had risked everything for American handlers were left behind in the chaotic evacuation, many ultimately facing execution or re-education camps. After the 2021 withdrawal from Afghanistan, similar scenes played out again.

These weren't failures of intention but of moral foresight—the inability to fully calculate the human costs of intelligence relationships.

Yet history also offers counterexamples of moral courage. CIA officer Gust Avrakotos defied his superiors to continue supporting Afghan fighters when policy shifted. MI6's Michael Dwyer resigned rather than participate in operations he considered unethical during the Troubles in Northern Ireland. NSA whistleblower Thomas Drake sacrificed his career to expose surveillance programs he believed violated constitutional principles.

These cases remind us that while institutions may sometimes operate in amoral logic, individuals working for them retain their moral agency—often at great personal cost.

As declassified records continue to shed light on this ethical area, we gain not just operational knowledge but a deeper understanding of how democratic societies balance security concerns with core values. Intelligence history teaches us that this balance isn't achieved through abstract principles alone. Instead, it's preserved through the choices of individuals who decide, in specific moments, which lines they will cross and which they won't.

Conclusion

Immediately after pandemic travel restrictions lifted—when Italy's ancient sites stood momentarily free from crowds—I found myself standing among the ruins of a Roman courier station outside Pompeii.

And, then it hit me....

I was walking through a 2,000-year-old intelligence outpost. Those weathered stone buildings once housed the messengers of the Cursus Publicus, the Roman postal service that doubled as its spy network. Can you imagine the secrets that passed through here? Emperors made world-changing decisions based on confidential information sent by vetted couriers across thousands of miles.

That's when I realized something fascinating about spy history. The tools keep dramatically changing, but the people remain remarkably the same. The Venetians running spy networks in the 1500s would immediately recognize the basic techniques used by CIA officers today. Elizabeth I's spymaster would nod knowingly at the challenges of running agents in hostile territory that modern intelligence services still face.

For history buffs like us, these connections across centuries aren't just interesting, they're actually useful in our daily lives. Think about it: Renaissance Venice protected state secrets by compartmentalizing knowledge and operating strictly on a need-to-know basis. Doesn't that sound like modern cybersecurity recommendations for businesses? They were dealing with handwritten notes instead of digital files, but the principles are identical.

The history of intelligence analysis is just as relevant. I love the story of William Friedman breaking Japan's Purple cipher before World War II. He used the same step-by-step approach that helped analysts track down Osama bin Laden decades later. I've started applying these same techniques when trying to figure out whether a news story is legitimate or misleading.

Have you noticed how much harder it's gotten to separate fact from fiction these days? I've found myself thinking like a Cold War counterintelligence officer when scrolling through my news feed. Why is this particular story being pushed now? Who benefits from this interpretation? What important context might be missing? These questions come straight from intelligence tradecraft, and they've never been more valuable.

The personal security practices of historical spies are surprisingly practical, too. Queen Elizabeth's spymaster, Walsingham, survived in incredibly dangerous times by carefully controlling what information he shared. He kept separate networks of contacts and documented important interactions. Sounds exactly like the advice cybersecurity experts give us today: use different passwords, be careful what you share online, and keep records of suspicious contacts.

What really fascinates me about intelligence history is how it sheds light on the constant tension between security and liberty in free societies. Remember the controversy over the Zimmermann Telegram in 1917? Britain had to admit it was reading American diplomatic communications to prove that Germany was plotting against the U.S. The outrage and debate that followed mirror exactly what happened with the Snowden revelations about NSA surveillance. We've been wrestling with these same fundamental questions for centuries.

I think this historical perspective changes how we see our role as citizens. The Church Committee investigations of the 1970s showed how public pressure could dramatically reform intelligence practices that had crossed legal and ethical lines. Your voice actually matters in these debates. Just as it did when American colonists protested British searches without warrants, which directly influenced the Fourth Amendment protections we still rely on.

Despite all our amazing technology, intelligence work remains fundamentally human. One of my favorite spy stories is Operation Mincemeat from World War II. British intelligence used a corpse carrying fake invasion plans to completely fool Nazi Germany. Its success didn't depend on fancy technology, but on deep psychological insight into how German intelligence officers thought and reacted. It reminds me that understanding people, not mastering gadgets, remains the heart of both good intelligence work and everyday critical thinking.

Next time you visit your local history museum, try looking for the intelligence stories hiding in plain sight. That Revolutionary War battle they feature? It might have been won because of a farmer who reported enemy troop movements. Those diplomatic documents in the display case? They probably contain carefully crafted language designed to mislead foreign spies. The ordinary-looking merchant's ledger? It might conceal an extensive intelligence network that influenced national policy. These hidden stories are everywhere once you start looking for them.

I find it oddly reassuring how cyclical intelligence reforms have been throughout history. The pendulum between tighter security measures and stronger civil liberties protections has swung back and forth for centuries. Ancient Rome, Renaissance Venice, Elizabethan England, Revolutionary America—they all grappled with these same tensions. It helps me see today's debates as part of a continuing conversation rather than sudden crises.

Looking ahead, it's clear that AI, commercial satellite imagery, and massive data collection are transforming intelligence work as dramatically as the telegraph, radio, and computer did in previous eras. But history teaches us that while methods change constantly, principles endure. The questions that kept Sir Francis Walsingham up at night in the 1580s remain just as relevant today: How do we know what's true? Who can we trust? How do we protect what matters most?

So, as we finish this historical journey together, I hope you'll see these spy stories not just as entertaining tales but as valuable wisdom handed down through the centuries. The same skills that helped intelligence officers survive and succeed—critical thinking, healthy skepticism, close attention to detail, and understanding human behavior—are exactly what we need to navigate today's complex information environment.

The next time you double-check your phone's privacy settings or question whether a viral news story seems legitimate, remember you're applying lessons learned through centuries of intelligence history. That connection across time isn't just intellectually satisfying, it's genuinely useful in our everyday lives. After all, in a world full of information and disinformation, we're all spies now, sifting through the noise to find the signal.

Glossary

355: Code number for the female operative in the Culper Ring who gathered intelligence by attending British social events.

Advanced persistent threat (APT): Sophisticated, long-term cyber espionage operations attributed to state-sponsored groups, such as China's APT10, which conducted the Cloud Hopper campaign.

Aerial photography: Intelligence-gathering technique that became vital during World War II, allowing analysts to identify enemy installations, troop movements, and defensive positions from captured images.

Alan Turing: British mathematician who designed electromechanical "bombes" to break the German Enigma code at Bletchley Park, shortening World War II by years, only to be prosecuted for homosexuality in 1952 and subjected to chemical castration.

Albrecht von Wallenstein: Imperial commander during the Thirty Years' War who developed an extensive intelligence network that gave him significant military advantages.

Aldrich Ames: CIA counterintelligence officer who spied for the Soviet Union from 1985 to 1994, compromising numerous American intelligence operations and causing the execution of at least ten CIA sources in exchange for $4.6 million.

Alien Office: A British organization established during the Napoleonic Wars that monitored foreign visitors to identify potential French agents.

Aline Griffith, Countess of Romanones: American spy who maintained cover as a socialite in Madrid while running a network that exposed Nazi smuggling operations through neutral Spain.

Allen Dulles: Influential early director of the CIA who shaped the agency's identity and expanded its operations beyond intelligence gathering into covert action.

The Americans: Television series (2013–2018) that followed Soviet deep-cover operatives living as an American family in 1980s Washington, D.C., noted for its realistic portrayal of the psychological strain of maintaining false identities.

Anti-Bond: Term used to describe fictional spy characters like George Smiley who represent the antithesis of James Bond, emphasizing patience, psychological insight, and bureaucratic navigation rather than action and glamour.

Anthony Blunt: Member of the Cambridge Five spy ring who worked in British counterintelligence while passing information to the Soviets.

Argo: An American thriller film about a CIA agent who rescues six Americans in Tehran during the 1979 Iranian hostage crisis.

Arlington Hall: The U.S. Army's cryptanalytic headquarters during World War II where women made up approximately 80% of the workforce, though many were classified in lower professional categories despite doing work equal to male counterparts.

ARPANET: The U.S. Advanced Research Projects Agency Network, a communications system designed to survive a Soviet nuclear attack that later evolved into the foundation of the internet.

Arrow riders: Elite Mongol intelligence agents who carried arrow-shaped badges granting them priority access to the Yam relay system, enabling rapid intelligence transmission across the empire.

Austin Roe: A member of the Culper Ring who served as a courier, transporting intelligence through British checkpoints disguised as a merchant making deliveries.

Babington plot: A 1586 conspiracy to assassinate Queen Elizabeth I and replace her with Mary, Queen of Scots, uncovered by Walsingham's intelligence network through intercepted encoded letters.

Battle of Boju: A significant military engagement in 506 B.C.E., where Wu forces defeated the more powerful Chu state, largely due to intelligence provided by internal spies in the Chu court.

Battle of Megiddo: An important military conflict (c. 1500 B.C.E.) where Pharaoh Thutmose III defeated a coalition of Canaanite rulers, utilizing intelligence to choose an unexpected route and achieve tactical surprise.

Battle of Mohi: A decisive 1241 conflict where Mongol forces defeated Hungarian defenders, aided by extensive pre-battle intelligence gathered by arrow riders.

Benedict Arnold: American Revolutionary War general whose plot to surrender West Point to the British was exposed by intelligence operations, becoming a case study in counterintelligence failures.

Benjamin Tallmadge: The 26-year-old Yale graduate who built America's first organized intelligence network (the Culper Ring) during the Revolutionary War.

Berlin Tunnel: A joint CIA-MI6 operation in 1954 involving a quarter-mile tunnel under East Berlin that tapped into Soviet communication lines, yielding thousands of hours of recorded conversations despite being compromised by Soviet mole George Blake.

Betty Shinn: American codebreaker who helped break Japanese shipping codes, enabling U.S. submarines to sink millions of tons of enemy supplies.

Black propaganda: Disinformation tactics used by *shinobi* to spread carefully crafted rumors designed to sow distrust among enemy commanders.

Black site: Secret detention facility operated by the CIA outside U.S. territory where enhanced interrogation techniques were employed following 9/11.

Black Tom explosion: A 1916 act of German sabotage that destroyed a munitions depot in New Jersey, causing damage equivalent to a 5.5

magnitude earthquake and accelerating America's path toward entering World War I.

Bletchley Park: The British codebreaking center during World War II, where women constituted 75% of the workforce, contributing significantly to breaking German codes.

Blue Moon: Codename for the low-altitude surveillance missions flown by Navy RF-8A Crusader jets during the Cuban Missile Crisis, with pilots flying just 500 feet above Cuban soil to collect detailed imagery.

BULLRUN: NSA program revealed in the Snowden leaks that involved deliberately weakening encryption standards to facilitate surveillance.

Bureau of Investigation: The precursor to the FBI, which expanded rapidly after the Black Tom explosion to counter foreign espionage and sabotage threats.

Bureau of Statistics: Napoleon's intelligence service, deceptively named to conceal its true function of foreign espionage and information gathering.

Cabinet Noir ("Black Cabinet"): Cardinal Richelieu's secret intelligence team in France that intercepted communications by steaming open letters, breaking codes, and copying diplomatic messages.

Caesar Cipher: One of history's first systematic encryption methods, developed by Julius Caesar, where each letter in a message is shifted by a fixed number of positions in the alphabet.

Cambridge Five: Soviet spy ring that included high-ranking MI6 officer Kim Philby, whose betrayal influenced John le Carré's portrayal of moral ambiguity in espionage fiction and caused catastrophic damage to Western operations.

Castra Peregrina: The headquarters in Rome from which the Frumentarii operated their intelligence network throughout the Roman Empire.

Censorate (Yushi Tai): An ancient Chinese intelligence and oversight institution established during the Han Dynasty that gathered intelligence, monitored corruption, and assessed public opinion.

Center for Mission Diversity and Inclusion: Established by the CIA in 2010 to address historical biases and actively recruit from previously marginalized communities.

Chalk marks: Signals used by spies to communicate, such as leaving specific markings on mailboxes to indicate a dead drop was ready or a meeting was needed.

Church Committee: 1970s Senate committee that investigated CIA assassination plots and other intelligence abuses, leading to significant reforms and executive orders limiting certain activities.

Cipher symbols: Coded characters or marks used to encrypt messages, making them unreadable to anyone without the proper decryption key.

Cloud Hopper: A Chinese state-sponsored cyber espionage operation that compromised IT service providers to gain access to their clients' networks across 12 countries, stealing hundreds of gigabytes of industrial secrets and proprietary technologies.

Code talkers: Native American servicemen, primarily Navajo, who used their indigenous languages to create unbreakable military codes during World War II, despite earlier persecution for speaking these same languages.

Commercial cover: The practice of intelligence officers posing as businesspeople to operate in foreign countries without raising suspicion, a technique dating back to Renaissance espionage.

Committee of Public Safety: The de facto executive government during the Reign of Terror in the French Revolution that used mass surveillance to identify perceived enemies of the revolution.

Compression of time: Common distortion in spy films where operations that would realistically take months or years occur within days.

Coordinated inauthentic behavior: Term used by social media platforms to describe networks of accounts working together to mislead users about who they are and what they're doing, often as part of state-sponsored influence operations.

CORONA: The first successful American photo reconnaissance satellite program, operating from 1960 to 1972, which revolutionized intelligence gathering by providing imagery of previously inaccessible areas.

Counterintelligence: Security activities aimed at protecting an organization from espionage or sabotage by hostile intelligence services; Athens pioneered these operations.

Cousins relationship: The close intelligence partnership developed between MI6 and the CIA during the Cold War, involving sharing intelligence, coordinating operations, and jointly running agents.

Cribs: In cryptography, known or predictable content in encrypted messages that provides entry points for codebreakers, such as weather reports or standard military phrases.

Crypteia: Sparta's secret service that initially controlled their slave (helot) population but later expanded to intelligence gathering against Athens during the Peloponnesian War.

Culper Ring: The code name for the American spy network organized by Benjamin Tallmadge during the Revolutionary War that operated in British-occupied New York.

Dead drops: Secret locations where intelligence materials could be left by one operative and later retrieved by another, avoiding direct contact between agents.

Deepfake: AI-generated video and audio that can realistically mimic real people, creating opportunities for intelligence agencies to produce convincing but false content of political leaders or other targets.

Department V: A KGB specialized assassination unit formed in the 1960s, reportedly named after the fictional SMERSH from James Bond novels, demonstrating fiction's influence on actual intelligence structures.

Domesday Book: A comprehensive survey of England ordered by William the Conqueror in 1086 that documented land ownership and resources throughout the kingdom, serving as an intelligence asset.

Donald Maclean: British diplomat and member of the Cambridge Five who fled to Moscow in 1951 after his espionage activities for the Soviet Union were discovered.

Double agents: Individuals who appear to serve one side while actually working for the other; Walsingham specialized in turning captured Spanish agents to feed false information back to Spain; In Sun Tzu's system, operatives who appeared to work for the enemy while actually serving their master, considered both the most valuable and most dangerous type of spy.

Duke William of Normandy: Later known as William the Conqueror, he effectively used intelligence networks to avoid direct military confrontation with invaders in 1054 before eventually conquering England in 1066.

East StratCom Task Force: European Union organization established to detect and respond to Russian disinformation campaigns targeting European audiences.

Edward IV: A king during the Wars of the Roses who utilized superior intelligence networks to mobilize troops with remarkable speed and outmaneuver his enemies.

Eleanor of Aquitaine: A medieval queen who built an intelligence network primarily composed of women who could gather information while being underestimated due to gender prejudices.

Elizabeth Friedman: American cryptanalyst who broke enemy codes and testified against Nazi spies, helping convict 33 German agents despite being classified as "sub-professional.

End-to-end encryption: Modern secure communication method that prevents third parties from accessing data while it's transferred; conceptually evolved from Renaissance-era encryption techniques.

Enhanced interrogation techniques: Euphemism for controversial interrogation methods used after 9/11, including sleep deprivation, stress positions, and waterboarding, developed by psychologists James Mitchell and Bruce Jessen for $81 million.

Enigma machine: German encryption device with 159 million possible combinations that looked like an oversized typewriter but utilized rotating wheels and electrical connections to create codes the Germans believed unbreakable until Turing's work.

Expendable spies: Operatives sent on extremely dangerous missions with minimal chance of survival, often used to verify intelligence or test enemy counterintelligence capabilities.

Extraordinary rendition: Practice of transferring terrorism suspects to countries known to use torture for interrogation, allowing Western intelligence agencies to benefit from information obtained through methods they legally couldn't use themselves.

False flag operation: Actions designed to appear as though they were carried out by another country or group, such as cyberattacks containing planted "digital fingerprints" to mislead investigators about the true source.

Farewell Dossier: Collection of documents provided by KGB Colonel Vladimir Vetrov (codenamed "Farewell") that revealed Soviet technological espionage operations and led to a CIA counteroperation to feed flawed technology to Soviet agents.

Feng Zheng: Chinese imperial secret police who operated within hierarchical systems while maintaining operational flexibility.

Firehose of falsehood: Russian disinformation technique that floods media with numerous contradictory narratives about an event, not to convince people of any single explanation but to create confusion and doubt through sheer volume.

Five Eyes: Intelligence alliance between the United States, United Kingdom, Canada, Australia, and New Zealand that shares signals intelligence and has established protocols for handling ethical differences in operations.

Francis Walsingham: Queen Elizabeth I's Principal Secretary, who established what historians consider England's first professional intelligence service, pioneering systematic espionage methods.

Frequency analysis: A code-breaking technique where analysts examine how often certain symbols appear in encrypted text, matching them to the known frequency of letters in a language.

Frumentarii: Roman intelligence agents who began as grain collectors before evolving into the emperor's personal spy service.

Gallic acid: A key ingredient in the invisible ink used by Revolutionary War spies, derived from oak apples.

GCHQ (Government Communications Headquarters): Britain's signals intelligence agency that worked closely with the American NSA to intercept and analyze Soviet communications.

General Intelligence Directorate (GID): Jordan's intelligence service that became a valuable counterterrorism partner for Western agencies after 9/11, raising ethical questions about cooperation with agencies using prohibited interrogation techniques.

George Smiley: John le Carré's recurring character who represents a realistic intelligence officer focused on psychological understanding and patience rather than physical action.

Graham Greene: Former MI6 officer who worked under Kim Philby during World War II and whose novels like *The Quiet American* and *The Human Factor* explore the moral costs of espionage.

Great Paris Cipher: Napoleon's complex military code that was broken by British intelligence officer Captain Sir George Scovell, giving Wellington crucial intelligence advantages.

Guy Burgess: British diplomat and member of the Cambridge Five who defected to the Soviet Union with Donald Maclean in 1951.

Harriet Tubman: Beyond her work with the Underground Railroad, a Union intelligence operative who built spy networks in Confederate territory and led the Combahee Ferry Raid.

Hope Royal Commission: 1970s Australian investigation into intelligence service overreach that shaped Australia's stricter restrictions on domestic spying compared to some Five Eyes partners.

Hotel Terminus: Klaus Barbie's headquarters in Lyon, France, where he conducted brutal interrogations of French Resistance members; later repurposed as the Center for the History of the Resistance and Deportation.

House of Life: An ancient Egyptian institution that served dual purposes as both a religious temple and an early intelligence headquarters.

HUMINT: short for "human intelligence"; information collected from human sources, as opposed to technical means; the primary method of intelligence gathering in ancient times.

Hunt forward operations: U.S. Cyber Command activities conducted with allied nations to identify and counter foreign influence campaigns before they can target American audiences.

Iga and Koga: Japanese regions that became centers for *shinobi* training during the Sengoku period, producing operatives who sold their intelligence services to various feudal lords.

Influence operations: Strategic activities aimed at affecting the perceptions and behaviors of target audiences; a specialty of Byzantine intelligence.

Internal spies: Sun Tzu's category for operatives placed within an enemy's inner circle, providing critical information about command disputes and defensive weaknesses.

Internet Research Agency (IRA): Russian organization established in St. Petersburg around 2013 that employed hundreds of people to create and manage thousands of fake American social media accounts, organizing at least 129 real-world events during the 2016 U.S. election period.

James Angleton: CIA counterintelligence chief whose close friendship with Kim Philby led to devastating intelligence leaks and whose subsequent hunt for Soviet moles created a "wilderness of mirrors" that paralyzed American intelligence operations.

James Lafayette: An enslaved man who served as a spy for the Americans during the Revolutionary War, providing crucial intelligence that contributed to victory at Yorktown.

Jean Moulin: The highest-ranking member of the French Resistance who was arrested and tortured by Klaus Barbie in 1943, becoming a martyr for the resistance after dying from his injuries.

Jerry Whitworth: U.S. Navy communications specialist recruited by John Walker Jr. to join his spy ring, providing classified cryptographic information to the Soviets.

John le Carré: Pen name of David Cornwell, a former MI5 and MI6 officer whose novels like *The Spy Who Came in From the Cold* and *Tinker Tailor Soldier Spy* presented intelligence work as morally ambiguous and bureaucratically constrained.

John Walker Jr.: Former Navy communications specialist who ran a family spy ring from 1967 to 1985, giving the Soviets the ability to decode Navy messages and recruiting his son Michael, brother Arthur, and friend Jerry Whitworth.

KGB: The Soviet intelligence agency formed in 1954 that combined foreign intelligence, counterintelligence, internal security, and border guard functions under a single organization.

Kim Philby: British intelligence officer who spied for the Soviet Union from the 1930s until his 1963 defection to Moscow, compromising countless Anglo-American operations while heading MI6's Soviet counterintelligence section.

Klaus Barbie: Nazi Gestapo chief in Lyon known as the "Butcher of Lyon" who tortured resistance fighters and orchestrated the deportation of Jewish children; later worked for U.S. intelligence before being brought to justice.

Knights Templar: A medieval military order that developed an extensive international intelligence network disguised within their banking and religious activities.

Law of Suspects: French Revolutionary legislation that allowed for the arrest of anyone showing insufficient enthusiasm for the revolution, used to justify mass surveillance.

Leak and amplify: Intelligence tactic where agencies hack sensitive materials, selectively leak them, then use social media networks to maximize their exposure and impact, as seen in the 2016 DNC hack.

Line X: The KGB scientific and technological espionage program that systematically stole Western technology rather than developing original Soviet innovations.

Listening posts: Hidden chambers or passages built into castle architecture, specifically designed for eavesdropping on conversations.

The Lives of Others: 2006 German film depicting Stasi surveillance in East Germany, praised for its realistic portrayal of the methodical nature of surveillance work and its psychological effects.

Local spy: In Sun Tzu's classification, an agent who blends into border communities to study terrain, identify defensive gaps, and gauge public sentiment in enemy territories.

Marie-Madeleine Fourcade: Leader of a French resistance network who provided intelligence mapping German defensive positions before D-Day and once escaped prison by removing her clothes to squeeze between cell bars.

Mata Hari: Dutch exotic dancer (born Margaretha Zelle) executed for espionage in 1917, whose actual intelligence value was minimal but whose execution became a powerful propaganda symbol.

Mizugumo: A "water spider" device created by *shinobi* for crossing moats and water barriers, essentially primitive snowshoes that distributed weight evenly.

Moral ambiguity: A central theme in realistic spy fiction that acknowledges the ethical compromises and unclear distinctions between "good" and "evil" in actual intelligence work.

Moral injury: Psychological harm resulting from perpetrating, failing to prevent, or witnessing acts that violate one's moral beliefs, experienced by many intelligence officers involved in controversial programs.

Mortimer's Hole: A secret tunnel in Nottingham Castle used by spies to enter and exit undetected, later famous for the capture of Roger Mortimer in 1330.

Nathan Hale: American spy executed by the British in 1776, famous for his last words: "I only regret that I have but one life to lose for my country" (*Rev. War Biography: Nathan Hale*, 2019).

National Photographic Interpretation Center: CIA facility where analysts identified Soviet missiles in Cuba from U-2 reconnaissance photographs in October 1962.

National Security Act of 1947: Legislation signed by President Truman that created the CIA as the first permanent American peacetime intelligence service.

Noor Inayat Khan: British SOE agent of Indian descent who served as the last remaining radio operator in Nazi-occupied Paris after her network was compromised, refusing to reveal information despite torture and eventually being executed at Dachau concentration camp.

NSA (National Security Agency): The U.S. intelligence organization responsible for global monitoring, collection, and processing of information and data for foreign and domestic intelligence and counterintelligence purposes.

Oleg Penkovsky (codenamed "TOUCHDOWN"): Soviet military intelligence officer who provided crucial information to the West during the Cuban Missile Crisis, later executed for his espionage.

Olympic Games: Codename for the joint American-Israeli operation that created and deployed the Stuxnet worm against Iran's nuclear facilities.

Omne Datum Optimum: A papal bull that granted the Knights Templar exceptional autonomy, enabling them to operate intelligence networks independent of local rulers.

Operation FAREWELL: 1982 CIA operation that fed defective computer code to Soviet spies, resulting in a three-kiloton explosion in a Siberian gas pipeline without any American agents on Russian soil.

Operation Fortitude: An Allied deception operation before D-Day that convinced German high command that the main invasion would come at Calais rather than Normandy.

Pearl Witherington: Female SOE agent who took command of a 3,500-strong French Resistance network after her male colleague was captured, becoming so effective that the Germans placed a million-franc bounty on her head.

Peloponnesian War: The protracted conflict (431–404 B.C.E.) between Athens and Sparta that saw significant developments in espionage and counterintelligence techniques.

The Prince: Niccolò Machiavelli's famous political treatise that, beyond its overt political philosophy, contained principles applicable to intelligence gathering.

Principal Secretary: A high government position in Tudor England that combined the roles of chief minister, head of intelligence, and secretary of state.

PRISM: NSA surveillance program revealed by Edward Snowden in 2013 that collected data directly from the servers of major internet companies, including Google, Facebook, Microsoft, and Apple.

Q Branch: The fictional technical division in James Bond films that became so culturally iconic it influenced how actual intelligence technical services approached their mission.

RAF Menwith Hill: A British base housing massive receivers that intercepted Soviet military communications from thousands of miles away.

Radio direction finding: Technology used to locate enemy transmitters by triangulating the source of radio signals, providing valuable intelligence on enemy positions.

Ratline: Nazi escape routes through Italy and other countries that helped war criminals like Klaus Barbie flee to South America after World War II, often operated with the help of Catholic clergy.

Reign of Terror: Period during the French Revolution (1793–1794) characterized by mass executions of "enemies of the revolution," many identified through the Committee's surveillance network.

Robert Townsend (Culper Jr.): A key member of the Culper Ring who gathered intelligence from British officers while working as a merchant in British-occupied New York.

Roger de Flor: A former Templar admiral who developed an extensive Mediterranean intelligence network with agents disguised as traders in port cities.

Room 40: British Admiralty's codebreaking operation during World War I that successfully decoded German naval communications, including the Zimmermann Telegram.

Sack of Magdeburg: A devastating 1631 military event during the Thirty Years' War, where Catholic forces destroyed the Protestant city of Magdeburg, largely due to intelligence failures.

Scytale: An early Spartan cryptographic tool using a wooden rod around which a strip of parchment was wrapped spirally, allowing messages to be encrypted and decrypted.

Section 215: Provision of the USA PATRIOT Act used to justify bulk collection of American telephone metadata until reformed by the USA FREEDOM Act.

Sengoku period: The era of Japanese history (1467–1615) marked by political fragmentation and warfare among competing feudal lords, creating ideal conditions for specialized intelligence services.

SERE (Survival, Evasion, Resistance, Escape): Military program whose techniques were adapted for enhanced interrogation, despite being originally designed to train American personnel to resist torture if captured.

Sherman Kent School: CIA training facility where intelligence officers receive ethics education including historical case studies and scenario-based exercises on complex moral dilemmas.

Shinobi shōzoku: The black suit associated with ninja in popular culture, though historical shinobi more commonly used disguises like merchants, monks, or performers.

SMERSH: Fictional Soviet counterintelligence organization in Ian Fleming's James Bond novels, derived from the Russian phrase "SMERt' SHpionam" (Death to Spies).

Source reliability rating system: An evaluation method attributed to Machiavelli that assessed the trustworthiness and accuracy of different intelligence sources.

Source validation: The practice of verifying intelligence by comparing information from multiple sources, as exemplified by Henry V's interrogation techniques during the siege of Rouen.

Special Operations Executive (SOE): British organization formed to conduct espionage, sabotage, and reconnaissance in occupied territories during World War II, which recruited numerous women for dangerous field operations.

Stinger missiles (FIM-92): American-made portable anti-aircraft weapons supplied covertly to Afghan mujahideen fighters that neutralized Soviet air superiority and contributed significantly to the Soviet withdrawal from Afghanistan.

Strategic deception: Military tactics involving deliberate misinformation or concealment to confuse enemies, often based on intelligence gathering.

Stuxnet: Highly sophisticated computer worm discovered in 2010 that specifically targeted Iran's nuclear centrifuges by taking control of Siemens industrial controllers while hiding its presence, destroying nearly 1,000 centrifuges and delaying Iran's nuclear program by an estimated 18-24 months.

Substitution cipher: An encryption method where each letter is replaced with another symbol, character, or letter; the Caesar Cipher is an example.

SVR: The Russian foreign intelligence service created after the dissolution of the KGB following the collapse of the Soviet Union in 1991.

Technical Services Division: CIA department responsible for developing specialized equipment and techniques for intelligence operations.

Teutoburg Forest: Site of a devastating Roman military defeat where Germanic tribes ambushed General Varus and three legions, partially due to intelligence failures.

The Thing: A passive listening device concealed inside a wooden Great Seal of the United States, given to the American ambassador in Moscow in 1956, which activated without batteries when targeted by specific radio frequencies.

The Thirty Years' War: A major European conflict (1618–1648) that saw significant developments in intelligence operations, particularly through Cardinal Richelieu's networks.

Thutmose III: Egyptian pharaoh (often called the "Egyptian Napoleon") who developed one of the world's first documented spy networks around 1500 B.C.E.

Tinker Tailor Soldier Spy: John le Carré novel and 2011 film adaptation that presents intelligence work as primarily intellectual rather than action-oriented.

Tony Mendez: CIA disguise specialist who created the fake film production cover story for the extraction of American diplomats from Iran, as portrayed in *Argo*.

Tradecraft: The specialized skills, methods, and techniques used by intelligence operatives to conduct espionage operations covertly and effectively.

Transposition cipher: An encryption technique that rearranges the positions of characters rather than replacing them; the scytale was an early example.

Triangulating information: The method of comparing reports from multiple sources to separate fact from fiction and establish reliable intelligence.

Twenty Committee (XX Committee): British counterintelligence operation during World War II that turned captured German agents into double agents who fed false information back to Germany.

Ultra: The codename for intelligence derived from decrypted high-level German communications during World War II, considered so valuable that military commanders sometimes had to allow attacks to proceed to protect the secret of its existence.

Ultra program: Allied intelligence operation during World War II that successfully broke German codes, demonstrating the evolution of intelligence techniques from Renaissance foundations.

UPSTREAM collection: NSA program that intercepted data flowing through the internet's physical infrastructure.

USA FREEDOM Act: 2015 legislation that placed increased limitations on bulk data collection following the Snowden revelations, though many surveillance capabilities remained intact.

Virginia Hall: American spy with a prosthetic leg (nicknamed "Cuthbert") who became so effective in occupied France that the Gestapo launched a nationwide manhunt to capture "the limping lady."

War council: A meeting of military leaders to discuss strategy; often targeted by early intelligence operations for infiltration or eavesdropping.

Warring States period: The era in ancient Chinese history (475–221 B.C.E.) characterized by intense competition among states, during which sophisticated intelligence practices developed.

Wars of the Roses: A series of English civil wars (1455–1487) during which intelligence networks played crucial roles in military strategy and dynastic struggles.

Water-holing: Technique where intelligence agencies compromise websites frequently visited by employees of target organizations to deliver malware, using social media analysis to identify suitable sites.

XKeyscore: NSA system revealed in the Snowden leaks that allows analysts to search through billions of records using just a person's name, email, IP address, or other identifier.

The Yam: The Mongol relay system that enabled rapid communication across their vast empire, with stations spaced approximately 20 miles apart providing fresh horses.

Yuri Andropov: KGB chairman who eventually became the leader of the Soviet Union, demonstrating the political power and influence of the intelligence agency within the Soviet system.

Zero Dark Thirty: 2012 film about the hunt for Osama bin Laden that sparked debate about whether intelligence officials compromised security by cooperating with filmmakers.

Zimmermann Telegram: A coded message from German Foreign Minister Arthur Zimmermann proposing an alliance between Germany and Mexico against the United States, intercepted and decoded by Room 40 in 1917, helping bring America into World War I.

References

Bleyer, B. (2022, March 21). *The myth of Agent 355, the woman spy who supposedly helped win the Revolutionary War.* Smithsonian Magazine. https://www.smithsonianmag.com/history/the-myth-of-agent-355-the-woman-spy-who-supposedly-helped-win-the-revolutionary-war-180979748/

Bush, George W. (2001, September 12). *Remarks by the president.* Archives.gov. https://georgewbush-whitehouse.archives.gov/news/releases/2001/09/print/20010912-6.html

Case study: Espionage. (n.d.). CDSE: Center for Development of Security Excellence. https://www.cdse.edu/Portals/124/Documents/casestudies/case-study-ames.pdf

Chinese state-led "Cloud Hopper" campaign targets technology service providers. (2024). Alliance for Securing Democracy. https://securingdemocracy.gmfus.org/incident/chinese-state-led-cloud-hopper-campaign-targets-technology-service-providers/

Clews, E. (2014, January 6). *Torture and the impact of 24 on America after 9/11.* E-International Relations. https://www.e-ir.info/2014/01/06/torture-and-the-impact-of-24-on-america-after-911/

Committee sensitive—Russia investigation only 116th congress 1st session senate (u)report of the report 116-xx select committee on intelligence United States senate. (n.d.). U.S. Senate Committee on Intelligence.

https://www.intelligence.senate.gov/sites/default/files/docum
ents/Report_Volume2.pdf

Cox, H. (2018, November 28). *Cracking stuff: How Turing beat the Enigma.*
Science and Engineering.
https://www.mub.eps.manchester.ac.uk/science-
engineering/2018/11/28/cracking-stuff-how-turing-beat-the-
enigma/

The Culper code book. (n.d.). George Washington's Mount Vernon.
https://www.mountvernon.org/george-washington/the-
revolutionary-war/spying-and-espionage/the-culper-code-book

Dawsey, J. (2019, May 29). *Targeting the most vulnerable: Klaus Barbie and
the Izieu children's home.* The National WWII Museum.
https://www.nationalww2museum.org/war/articles/klaus-
barbie-izieu-childrens-home

Domesday book: What can we learn about England in the 11th century? (n.d.).
The National Archives.
https://www.nationalarchives.gov.uk/education/resources/do
mesday-book/

Dulles, A. W. (n.d.). Quote. In Staff (2019, June 28). *NRO artifacts at
Spy Museum answer 60 year old questions.* National Reconnaissance
Office. https://www.nro.gov/news-media-featured-
stories/news-media-archive/News-
Article/Article/1896066/nro-artifacts-at-spy-museum-answer-
60-year-old-questions/

ET Online. (2024, September 21). *Here's how CIA and Mossad disrupted
Iran's nuclear program using a computer virus.* The Economic Times.
https://m.economictimes.com/news/defence/heres-how-cia-
and-mossad-disrupted-irans-nuclear-program-using-a-
computer-virus/articleshow/113545993.cms

Evans, E. (2019, August 27). *10 things you (probably) didn't know about Henry V and the Battle of Agincourt.* HistoryExtra. https://www.historyextra.com/period/medieval/things-you-didnt-know-facts-henry-v-battle-agincourt-shakespeare-hundred-years-war-france/

Fessler, P. (2019, April 19). *Mueller report raises new questions about Russia's hacking targets in 2016.* NPR. https://www.npr.org/2019/04/19/714890832/mueller-report-raises-new-questions-about-russias-hacking-targets-in-2016

Greenwald, G. (2013, July 31). *XKeyscore: NSA tool collects "nearly everything a user does on the internet."* The Guardian. https://www.theguardian.com/world/2013/jul/31/nsa-top-secret-program-online-data

Greenwald, G., & MacAskill, E. (2013, June 7). *NSA Prism program taps in to user data of Apple, Google and others.* The Guardian. https://www.theguardian.com/world/2013/jun/06/us-tech-giants-nsa-data

Gruber, K. E. (2022, February 3). *Black spies of the American Revolution.* American Battlefield Trust. https://www.battlefields.org/learn/articles/black-spies-american-revolution

Hendrix, J. (2021, May 31). *Researchers reverse-engineer 2016 Texas protest organized by Russian Internet Research Agency.* Tech Policy Press. https://techpolicy.press/researchers-reverse-engineer-2016-texas-protest-organized-russian-internet-research-agency

Hersh, S. M. (2004, April 30). *Torture at Abu Ghraib.* The New Yorker. https://www.newyorker.com/magazine/2004/05/10/torture-at-abu-ghraib

Hewitt, S. (n.d.). *The Canadian-less Caper: The controversy surrounding Oscar-winning film Argo.* University of Birmingham.

https://www.birmingham.ac.uk/research/perspective/argo-opinion

Hill, B. (2024, September 9). *6 spy movie gadgets that became a reality.* SlashGear. https://www.slashgear.com/1659238/spy-movie-gadgets-became-reality/

How John le Carré reinvented the spy novel. (2020, December 24). CrimeReads. https://crimereads.com/how-john-le-carre-reinvented-the-spy-novel/

Human Rights First. (2016, February 10). *"Enhanced interrogation" explained.* https://humanrightsfirst.org/library/enhanced-interrogation-explained/

Internet history sourcebooks project: Ancient history. (2025). Fordham.edu. https://origin-rh.web.fordham.edu/Halsall/ancient/aelius-hadrian.asp

Jasper, C. (2024, June 2). *Britain's real life "Q-branch" joins US war on fentanyl with blimp fleet.* Yahoo Tech. https://www.yahoo.com/tech/britain-real-life-q-branch-160000163.html

Jureidini, B. (2024, December 12). *The aristocratic answer to James Bond? Inside the secret life of Aline Griffith, Countess of Romanones, who uncovered KGB moles with the Duchess of Windsor.* Tatler. https://www.tatler.com/article/aline-griffith-countess-romanones

Klein, C. (2021, March 15). *Josephine Baker's daring double life as a World War II spy.* HISTORY. https://www.history.com/articles/josephine-baker-world-war-ii-spy

le Carré, J. (1963). Quote. In JekyllnHyde. (2010, February 13). *What is your favorite Cold War movie?*. Docudharma. https://www.docudharma.com/tag/16589

Leung, R. (2004, April 27). *Abuse of Iraqi POWs by GIs probed.* CBS News. https://www.cbsnews.com/news/abuse-of-iraqi-pows-by-gis-probed/

loek_essers. (2014, January 29). *German government faces legal action over NSA spying.* InfoWorld. https://www.infoworld.com/article/2193166/german-government-faces-legal-action-over-nsa-spying-2.html

Macdonald, H., Levy, A., Muller, G., & Lopresti, L. (2023, March 6). *The Soviet spy films that inspired a young Vladimir Putin.* ABC News. https://www.abc.net.au/news/2023-03-07/vladimir-putin-fan-of-soviet-spy-films/101973078

Mark, J. (2017, July 24). *Thutmose III at the Battle of Megiddo.* World History Encyclopedia. https://www.worldhistory.org/article/1101/thutmose-iii-at-the-battle-of-megiddo/

Maranzani, B. (2013, May 31). *Harriet Tubman: 8 facts about the daring abolitionist.* HISTORY. https://www.history.com/news/harriet-tubman-facts-daring-raid

Militarising big tech: The rise of Silicon Valley's digital defence industry. (2023, February 7). Transnational Institute. https://www.tni.org/en/article/militarising-big-tech

Moran, C. R., & Hammond, A. (2021). Bringing the "social" in from the cold: Towards a social history of American intelligence. *Cambridge Review of International Affairs, 34*(5), 616–636. https://doi.org/10.1080/09557571.2021.1960796

Neven, T. (2018, March 2). *Virginia Hall: The limping lady*. USSOCOM: United States Special Operations Command. https://www.socom.mil/virginia-hall-the-limping-lady

New King James Bible. (n.d.). Bible Gateway. https://www.biblegateway.com/versions/New-King-James-Version-NKJV-Bible/ (Original work published 1979)

Nuclear power in Iran. (2024, May 3). World Nuclear Association. https://world-nuclear.org/information-library/country-profiles/countries-g-n/iran

Numbers 13 – spies are sent into Canaan. (2015, December 17). Enduring Word Bible Commentary. https://enduringword.com/bible-commentary/numbers-13/

Omne datum optimum. (n.d.). OPCCTS: The Knights Templar of North America. https://www.knighttemplar.org/single-post/2018/03/16/Omne-Datum-Optimum

Patton, G. (n.d.). *George Patton quotes*. BrainyQuote. https://www.brainyquote.com/quotes/george_s_patton_102496

Reuters. (2008, May 16). FBI assists with detainee cases, differs with CIA. *Reuters*. https://www.reuters.com/article/world/fbi-assists-with-detainee-cases-differs-with-cia-idUSN16441950/

Rev. War biography: Nathan Hale. (2019, June 4). American Battlefield Trust. https://www.battlefields.org/learn/biographies/nathan-hale

Rollins, J. W. (2023, November 7). *Armed drones: Evolution as a counterterrorism tool*. EveryCRSReport.com. https://www.everycrsreport.com/reports/IF12342.html

Ross, B. (2012, April). Krypteia: A form of ancient guerrilla warfare. *Grand Valley Journal of History*, *1*(2).

https://scholarworks.gvsu.edu/cgi/viewcontent.cgi?article=10
25&context=gvjh

Santhanam, L. (2014, December 9). *Two military psychologists were paid $81 million to develop the CIA's enhanced interrogation techniques*. PBS News. https://www.pbs.org/newshour/politics/came-idea-use-enhanced-interrogation-techniques

Secret passage. (2025). Kids on the Net. http://www.kidsonthenet.com/castle/sugar.html

Senate refuses to poison the USA FREEDOM Act. (2025, June 2). TechFreedom. https://techfreedom.org/senate-refuses-to-poison-the-usa-freedom-act/

Sweeney, R. (2016, September 13). *The Edward Snowden book that answers all your questions*. Early Bird Books. https://earlybirdbooks.com/the-edward-snowden-book-that-answers-all-your-questions

Two Chinese hackers associated with the ministry of state security charged with global computer intrusion campaigns targeting intellectual property and confidential business information. (2018, December 20). Archives: U.S. Department of Justice. https://www.justice.gov/archives/opa/pr/two-chinese-hackers-associated-ministry-state-security-charged-global-computer-intrusion

Tzu, S. (n.d.). *Sun Tzu quotes*. Goodreads. https://www.goodreads.com/quotes/103957-all-warfare-is-based-on-deception-hence-when-we-are

University of Toronto. (2010, April 8). *Researchers shed light on ancient Assyrian tablets*. Phys.org. https://phys.org/news/2010-04-ancient-assyrian-tablets.html

The Wars of the Roses. (n.d.). BBC Bitesize. https://www.bbc.co.uk/bitesize/articles/zyfr8p3

Washington, G. (1777, July 26). Quote. In *security awareness posters.* (2025). U.S. Department of Commerce Office of Security. https://www.wrc.noaa.gov/wrso/posters/Security_Awareness _Posters-i0005.htm

William I "The Conqueror" (r. 1066-1087). (2018, August 3). The Royal Family. https://www.royal.uk/william-the-conqueror

William Wallace. (2025). Cranntara.scot. https://cranntara.scot/wallace.htm

Image References

Dibbly. Prompt: *A Cold War-era Berlin street at night, dimly lit by streetlamps. A tense standoff unfolds—two men in trench coats stand near a graffiti-marked section of the Berlin Wall, one subtly handing off a document while keeping an eye on his surroundings. In the background, shadowy figures lurk, watching from dark alleys and parked cars. A Soviet KGB officer observes from a distance, while a CIA operative, partially hidden behind a newspaper stand, monitors the exchange. The atmosphere is thick with suspense, a world where every glance could signal betrayal or discovery. The image should be black and white with the exception of the graffiti on the Berlin wall which should be displayed in bright red.* [AI-generated image]. Dibbly. https://dibbly.com

Dibbly. Prompt: *A complex, high-contrast black and white illustration of a global surveillance or espionage network. A vintage world map dominates the background, overlaid with bright streams, schematics, and glowing network lines. Silhouetted figures in Victorian-era detective or secret agent attire (top hats, trench coats) are scattered across the scene, standing and observing screens, maps, and old-fashioned control panels. A bright red glowing globe or satellite above the map casts dramatic light rays. The aesthetic is noir, cinematic, and mysterious, evoking themes of secret societies, intelligence gathering, and global strategy. Moody lighting, high detail, dramatic shadows. Image should be black and white except for the red globe.* [AI-generated image]. Dibbly. https://dibbly.com

Dibbly. Prompt: *A crowded colonial tavern in Boston, circa 1775. Benjamin Tallmadge and other Continental Army intelligence operatives are strategically seated at various tables, blending in with merchants, sailors, and locals. Some are subtly passing coded notes, using invisible ink, or whispering critical information. British loyalists are also present, creating an atmosphere of hidden tension. Soft candlelight illuminates wooden tables, tankards of ale, and shadowy corners where secret communications occur. The clothing should be period-accurate—waistcoats, tricorn hats, rough working-class attire. Emphasize the sense of danger and intrigue, with characters listening carefully to conversations and watching each other with cautious, calculating eyes. The image should be black and white with the exception of a few notes on the various tables that should be displayed in bright red.* [AI-generated image]. Dibbly. https://dibbly.com

Dibbly. Prompt: *A dimly lit attic in Nazi-occupied Paris. Noor Inayat Khan, a young woman of Indian descent, sits at a small wooden table, fully visible, her determined face illuminated by the soft glow of a Morse code transmitter. Her dark hair is neatly tucked beneath a headscarf, and she wears a modest coat suited for blending in. Her eyes are focused, her expression tense yet resolute as she taps out a coded message. Behind her, a small, tattered map of Europe is pinned to the wall, with a marked line stretching from Paris to London. Through the cracked attic window, the rooftops of Paris stretch out, with the faint silhouette of the Eiffel Tower confirming the location. The atmosphere is one of secrecy and danger, as if the enemy could be just outside. The image should be black and white with the exception of the eiffel tower which should be displayed in bright red.* [AI-generated image]. Dibbly. https://dibbly.com

Dibbly. Prompt: *A dimly lit city street at night, with an old-fashioned red phone booth glowing under a flickering streetlamp. Inside the booth, a mysterious figure is speaking in hushed tones, unaware of the surveillance. Outside the booth, several men dressed in dark suits and overcoats are positioned strategically, listening in. One leans against the wall, subtly adjusting an earpiece, while another holds a small recording device. Their faces are tense, focused. Behind them, a brick wall prominently displays the Five Eyes alliance flags—U.S., UK, Canada, Australia, and New Zealand—either painted or projected onto the surface, symbolizing the intelligence cooperation at play. The scene exudes a cinematic spy-thriller atmosphere, blending classic espionage with modern geopolitical surveillance.* [AI-generated image]. Dibbly. https://dibbly.com

Dibbly. Prompt: *A dramatic high-contrast black and white illustration of a medieval castle's stone watchtower at dusk, viewed from a low angle perspective. In the foreground, elongated shadows of hidden figures stretch across ancient stonework. The tower features narrow arrow slits emitting faint light, while a hooded figure in dark robes stands partially concealed in an archway. Subtle details include a cloaked messenger slipping between shadows, clutching a bright red book (which is the only color besides black and white in the picture), and barely visible smoke signals rising from a distant tower. The composition emphasizes Gothic architecture with deep shadows and stark lighting, creating an atmosphere of secrecy and surveillance. Textural details highlight the rough castle stones and weathered wooden elements, while dramatic clouds loom overhead, adding to the mysterious ambiance. Style: high contrast charcoal/ink illustration with strong emphasis on light and shadow play.* [AI-generated image]. Dibbly. https://dibbly.com

Dibbly. Prompt: *A futuristic intelligence war room with multiple digital screens displaying real-time social media feeds, deepfake videos, and AI-generated news articles. Analysts in dark suits examine engagement metrics, tracking viral misinformation campaigns with data visualizations. A shadowy figure types on a laptop, launching a coordinated bot attack to amplify false narratives. In the background, a web of interconnected social media accounts reveals a network of influence operations, with flags of different nations subtly integrated into the interface. The image captures the dual nature of social media as both a tool for intelligence gathering and a battleground for digital deception. The image should be black and white with the exception of a minor few symbols that should be displayed in bright red.* [AI-generated image]. Dibbly. https://dibbly.com

Dibbly. Prompt: *A high-altitude U-2 spy plane flies over Cuba, capturing detailed reconnaissance photos of Soviet missile sites below. In a dimly lit intelligence room, U.S. analysts intensely examine black-and-white aerial images, pointing at missile installations hidden among dense foliage. A large map of Cuba is pinned on the wall, marked with red circles indicating missile locations. In the background, military officials in Washington, D.C., including President Kennedy, stand in a tense briefing, while Soviet officers in Cuba scramble to conceal their missile launch sites under camouflage netting. The atmosphere is thick with Cold War tension, symbolizing the critical intelligence operations that shaped global history.* [AI-generated image]. Dibbly. https://dibbly.com

Dibbly. Prompt: *A high-energy behind-the-scenes shot of a modern spy movie in production. The set features a sleek, Cold War-era espionage scene, but with a James Bond twist—an Aston Martin spy car, its headlights firing concealed machine guns. A dashing secret agent stands nearby in a sharply tailored suit, exuding confidence, while a stunning woman in an elegant bright red dress watches the action unfold with intrigue. The director, wearing a headset, passionately directs the scene beside a professional camera rig. Crew members adjust lights and cameras, while vintage surveillance equipment decorates the set, adding to the tension of the unfolding spy drama. Bright studio lights illuminate the action, blending classic espionage aesthetics with modern filmmaking magic. The image should be black and white with the exception of the bright red dress.* [AI-generated image]. Dibbly. https://dibbly.com

Dibbly. Prompt: *A panoramic, historical intelligence scene blending ancient civilizations. In the foreground, Egyptian and Assyrian spies in period-*

accurate clothing examine clay tablets and scrolls. On the left, Thutmose III peers through a reconnaissance map. On the right, Spartan and Athenian intelligence operatives communicate covertly. The background shows a sweeping landscape with ancient city-states, trade routes, and military encampments. The image be in black and white, but a pyramid far back in the distance should be displayed in bright red, with intricate details showing maps, communication tools, and the strategic positioning of intelligence gatherers. Emphasize the sense of secrecy, strategic planning, and the subtle power of information gathering in ancient warfare. The composition should feel like a strategic war room merged with a historical tapestry, highlighting the sophisticated nature of early intelligence networks. [AI-generated image]. Dibbly. https://dibbly.com

Dibbly. Prompt: *A portrait of Genghis Khan sitting on a throne or elevated platform, surrounded by multiple intelligence agents and scouts reporting to him. Khan should be wearing traditional Mongol royal attire - fur-trimmed deel (robe), with intricate leather and metal details, looking powerful and contemplative. Around him, diverse agents display maps, whisper reports, and demonstrate communication methods - some dressed as merchants, others as scouts, representing his extensive intelligence network. The background should show a map of the vast Mongol Empire with strategic routes and territories subtly indicated. The lighting should be dramatic, emphasizing Khan's central role in gathering and using intelligence. It should be a black and white image except for a few minor details in the Genghis Khan attire (like jewelry) that should be bright red - to represent the Mongol aesthetic. Capture an atmosphere of strategic intensity, showing how information was the true power behind his conquests. The image should convey both the physical presence of Genghis Khan and the sophisticated intelligence apparatus he commanded.* [AI-generated image]. Dibbly. https://dibbly.com

Dibbly. Prompt: *A sophisticated split-composition black and white image divided between a Renaissance study and an Elizabethan war room. On the left, Machiavelli sits at an ornate desk surrounded by coded diplomatic dispatches, maps, and intercepted letters, his face illuminated by candlelight as he writes in his characteristic secretive style. The right side shows Walsingham's intelligence operation, with codebreakers working at tables covered in ciphered messages while agents report through a shadowy doorway. Connecting these scenes, a central spiral staircase leads upward (with its banister displayed in bright red), its steps formed from documents and coded messages, symbolizing the evolution of intelligence practices. The architectural details blend Italian Renaissance and Tudor Gothic styles. Strong chiaroscuro lighting creates dramatic shadows that suggest secrecy and intrigue. Hidden details include spy symbols carved into woodwork, secret compartments in walls, and reflections of informants in mirrors. The style emphasizes fine line work for architectural precision while using softer, more atmospheric techniques for the human elements, creating a sense of both technical precision and human intrigue.* [AI-generated image]. Dibbly. https://dibbly.com

Biography

Dexter Ingram started his national security career as a Naval Flight Officer before entering the high-stakes world of counterterrorism strategy. His State Department experience includes serving as the Director of the Office of Countering Violent Extremism, Acting Director for the Office of the Special Envoy to Defeat ISIS, and Deputy Director for the Office of Preventing WMD Terrorism.

Ingram's field experience spans multiple continents - from coordinating counterterrorism efforts at INTERPOL headquarters in Lyon, France; to leading U.S. interagency delegations to Asia and Africa focusing on nuclear proliferation deterrence; to serving as Senior Political Advisor in Afghanistan's volatile Helmand Province, as well as playing key liaison roles with both the FBI and DHS.

An avid collector of intelligence history artifacts, Ingram has assembled one of the most impressive private spy collections outside government archives. He currently serves on the boards of the International Spy Museum, Real Spy Comics, and the National Counterterrorism Innovation, Technology, and Education Center (NCITE).

In 2023, Ingram founded IN Network, a nonprofit dedicated to mentoring promising young minds of all backgrounds interested in careers in national security (in-network.org).

Subscribe to Ingram's Substack, Codename: Citizen, for insights and behind-the-scenes perspectives on national security, espionage, and public service and connect with him on LinkedIn at linkedin.com/in/dexteringram.

Thank You for Reading

I'm grateful you chose to spend your time with THE SPY ARCHIVE. Every reader brings these stories to life in their own way, and your perspective matters.

If you enjoyed the book - or even if there's something you think could make it better - I'd love to hear from you. Your review helps other readers discover the book and supports the mission behind it.

Scan the QR code to leave your thoughts.

All proceeds from this book go to IN-Network.org, a nonprofit dedicated to inspiring and guiding the next generation of national security leaders.

Thank you for being part of this journey and helping us make a difference.